Cory,

all the best in what
you
are called to do.

Scott

Mindful Ecology

Mindful Ecology
A Way Forward

by
Scott Erway

Griffin Creek Press

Mindful Ecology: A Way Forward
Published by Griffin Creek Press
Copyright © 2018 by Scott Erway

First Edition
This book is set in Alegreya Type Text

ISBN-13:978-0-9996696-4-8
ISBN-10:0-9996696-4-8

Griffin Creek Press
P.O. Box 190
Carnation,Wa. 98014

Dedicated, with love, to my children
Clare, Alexander and Michael.

Contents

Meditation Advice

Mindful Ecology Contemplations

Perspective and Peace of Mind

Introduction

Have you ever read a book or seen a film about the ecological crisis and felt overwhelmed by what you had been exposed to? The extent and speed with which our societies are remaking the earth is unprecedented. If even a tenth of the forecasts from ecology's models come to pass, the day after tomorrow is is almost too terrifying to think about.

Mindful Ecology is a way to think about these things.

This book has been written to serve as an invitation to meditation for those hurt by their awareness of the ecological crisis and not sure how to deal with it so it does not sour the whole of their lives. It is also written for those who have been following a meditative path for years and are looking to include elements of system science and ecology into their contemplations.

Most of us are never taught how to go about thinking slowly and carefully about things, nor are we taught to include an awareness of how our bodies are reacting to what we are thinking about. The result is that most of our cognitions remains little more than factoids, toys of the intellect, instead of becoming truths about our world we deeply and immediately understand. Around such truths we can form meaningful lives.

Mindful ecology asks if you are ready to take a radical step, one proportionate to the crisis of unsustainability we find ourselves in. We are in need of courageous people who can take the fight to the

Mindful Ecology

monster in our midst: the collapse of fossil fueled industrial civilization. We are in need of people trained to perform open soul surgery under triage conditions to aid those traumatized by the monster.

More and more people are waking up to the horror of the ecological crisis unfolding at a rapid pace throughout the earth. When the horror penetrates the heart - the mind is left numb. What, we wonder, are we to do?

Mindful ecology is one way of responding. It involves developing a direct relationship with the deepest issues. It seeks a profound understanding of the planetary sciences through contemplating them, thinking deeply about them, allowing them to make an impact on one's emotions and values.

I would like to take this opportunity to publicly thank the authors who were most influential in educating me, and provoking me to respond. Mindful Ecology is my heart felt thank you for the courage, integrity and honesty I found in their work. I am not sure it can be rightly understood apart from it. I have included a select bibliography but wanted to call out these powerful works for particular attention.

The best minds share seemingly simple models with us and speak only that which is most obvious, once it has been said. It is when we realize they stand alone in their trail-blazing that we recognize the true extent of the genius involved. William Catton's *Overshoot: The Ecological Basis of Revolutionary Change* introduces the key ecological findings pertinent to our actual circumstances in ideas and terms that allow us to think more clearly. Such service is invaluable. Here, in summary, is the story of the ecological message of our time according to Dr. Catton. The oil fueled industrial economy is a detritus feeder, subject to overshoot. The state of overshoot is sustained as long as the phantom acreage on which it relies remains available. In the Age of Oil the detritus feeder found just the food it needed to grow enormous, even giant, life-threateningly giant. Homo Colossus was born, the prosthetic extensions of our human reach through the power of the technology we strapped on our backs. Its giantism does not respect the limits which must be inherent to all things on our finite earth: its need to grow endlessly is destroying the biosphere. The death of Homo Colossus will likely be accompanied by a population die off, as is the way with detritus feeders. For those with ears to hear...

The basic statement on peak oil presented in Richard Hienberg's *The Party's Over: Oil, War, and the Fate of Industrial Societies* remains a solid accounting of an ecologist persuaded by the Hubbert Curve. There remains no rebuttal to the basic argument it presents, which is a nail in the coffin of Homo Colossus.

John Michael Greer corrects the pervasive dismissal of the real future we are making for ourselves in *The Long Descent, a Users Guide to the End of Industrialization* and *The Ecotechnic Future: Envisioning a Post Peak World*. Together they provide a point of view carefully leavened by historical precedent. It is a corrective to so much that is blindly taken for granted by a society equally smitten by visions of endless technological progress and cosmic sized apocalyptic fears. These books remind us the future we are going to get is the one we are making, which is by the way, much more frightening.

One film makes this short list, Timothy S. Bennett's *What a Way to Go, Life at the End of Empire* produced by Sally Erickson. At the heart of this film is a gut-wrenchingly honest soliloquy from an individual which had been deeply touched by the sadness and madness of what industrialized civilization is doing to the earth. It remains the classic rhetorical presentation, pleading with each of us to wake up and change our ways.

Finally, Derek Jensen's work has had a very powerful effect on me. *In A Language Older than Words* I heard a full-throated voice, powerful with an honest integrity determined to have its say. The scream, the revulsion, the human refusal to take the bully lying down or cover up their crimes; all these things speak in his work. They are dark works, flint for the soul in a dark night. They challenged me to do the same. For those to whom they resonate they bring the fully

Mindful Ecology

embodied human being onto the front lines to confront some hard truths. Homo Colossus abuses the earth. In the final analysis its toxicity was born from us, from our abuses of one another.

This shows us a way forward.

"... the plight of the unwanted child became potentially everyone's plight."

Overshoot, William Catton

1.
What is Mindful Ecology?

A Few Moments of Concern for the Earth

There is a serious threat to the well-being of numerous ecosystems on which human life depends. There is a serious threat to the well-being of the human societies we have built on unsustainable production technologies. These very real threats are the unintended consequences of our lifestyles. This is the ecological age, the time when the limits to growth are remaking the world, one day at a time.

We are not free to choose whether or not these changes will come to pass. We are free to choose how we meet them; free to decide, as individuals, how we will respond. Once educated into the reality of the harm we are causing the conscience insists we act. Nothing less honors the power of the visceral and emotional response the planetary tragedy evokes. What Mindful Ecology would like to suggest is that there is a very important work going on when we educate our minds about ecological realities and allow our hearts, in turn, to feel compassion for the earth and all sentient beings who call her home. This inner work with our own culturally influenced consciousness is, in my mind, the primary work needed in our time. As I read the tea leaves, the ecological crisis is a reflection of a crisis in our conception of ourselves as human beings. Contemplating ecological relationships teaches us directly where greed and hubris have used our ignorance against us. It shows us, with an immediacy hard to miss, what a more sustainable lifestyle will necessarily entail. I suggest we are ill, and that one cure consists of refusing to let fear dominate our minds by repressing what we know to be true about the actual ecological predicament the human race is in. To fight that repression does not take magical ceremonies or special prayers, nothing so esoteric is needed. We just need to learn to care enough to set aside a few moments of concern for the earth, ideally every day. Set aside a few moments to sit quietly and just slowly turn over in your mind's eye the wonders of nature, our nature, the good the bad and the ugly. Surely it is the least we can do.

Who knows, maybe letting these hard truths change us will ultimately be of benefit to the earth. I am sure it will be of benefit to ourselves. We play mind games to cover up the facts, playing pretend and make-believe with ourselves even while we know such an approach is likely to fail miserably. Facts are stubborn things.

Conversations about the ecological crisis typically are constrained to questions of political will and economic costs. This point of view assumes the biosphere would adapt to our self-defined needs, instead of things being the other way around. In contrast, the approach being sketched out here takes planetary deep time as primary and the people alive right now as but its current manifestation. By transplanting the context by which we understand our role as Homo Sapeins from the high stress, high stimulus, quarterly-profits driven environments of industrial consumerism, to repositioning ourselves within the long trends of evolutionary deep time, we have an opportunity to gain a perspective which grants an unassailable peace of mind. It also nurtures a profound commitment to the continuation of the human experience in all its diversity. The untrained mind fears highly improbable worst case scenarios for how this ecological crisis will play out. Part of the danger of repression is how monsters grow in the dark. Yes, times are hard and going to get a whole lot harder. No, this is not the end of the world. Yes, it is the end of fossil fueled giant sized industrialized civilization. No, that is not the end of cultural evolution full stop. Yes, there will be a population decrease for the human race with a high probability that the process will include episodes of heart-rending die offs. No, this will not remove all laughter and joy from the day-to-day life of our progeny, nor the sweet trembling terror of falling in love among our young.

Our societies have, by and large, behaved childishly by refusing to have mature discussions about the ecological facts. How can we begin to have an adult conversation, one in keeping with the magnitude of the subject ecology reveals? My suggestion is that there is no more direct route to these conversations we need to be having than by encouraging change in individuals. Individuals must first educate their mind about the reality of our ecological position. But this is not enough. The second step is to then contemplate what you have learned; to sit with quiet mind and still body and allow what you have learned to settle deeply into what you understand about the reality of the life you are living, this life we share equally with all other sentient beings. In these contemplations we allow ourselves to feel the emotional implications of what we know in our minds to be true. We allow the mind to teach the heart what is real. The result is that the heart teaches the mind what it must do to begin to address the wound that is tearing our civilization apart.

This is a time in which a loyalty focused directly on the planet's well-being is called for. Our sense of caring and responsibility runs naturally through our family and friends and out into our communities. All that is needed is to add the conscious appreciation of our place in the bigger scheme of things by including the land, water, and atmosphere of the places in which the lives of all these communities are played out.

If, as I propose, our institutional inheritance is proving incapable of adapting to the changing ecological reality of the twenty-first century, the only avenue left to explore is that which individuals can adopt. The model I work with is that each of our individual minds participates in the issues and psychology of the times. Each of us participates in the history of ideas. Each of us bears a shard of the darkness that is haunting our societies toying with fantasies of collective suicide in nuclear conflagrations. In the individual, the collective is working out its issues. Or is it that the collective is nothing more than the sum of individuals working out their issues? Either way, the point remains that *what we do with our own perceptions and emotions matters a great deal.*

What Is Mindful Ecology?

Those who can approach the challenges of the times with a cheerful heart are the real protectors of the earth; they are carrying the fight right to the enemy. There is an overwhelming sadness haunting the world these days. It feels dreary and old and everywhere there is fear that as a species we may not be able to control ourselves from falling once again into world wars, or worse. More and more it is becoming unusual to encounter people who are happy with a happiness that comes from deep within; happy just to be alive. Don't be fooled by the smiley-face, happy-dance charade of the consumer culture's advertised persona. That is nothing more than a mask we all adopt to one degree or another to get by. What I am talking about is the deep down assessment of life where what you really believe forms your character. Do you think this human experience is a good one or not? We are quick to say 'yes, of course, sure I do' - but our actions speak otherwise.

An enormous effort has gone into the making of the industrial infrastructure by which humanity's unprecedented population numbers are, for the most part, fed, housed and clothed. We rightly feel proud of our cleverness and accomplishments. It is indeed worth spending hours contemplating just how fortunate we are that the mind we experience is able to penetrate the mysteries of our environment so well. We take the ability to reason for granted all too easily. Living in this age, we are the beneficiaries of countless man-hours of research, experimentation and engineering creativity. Through the ability to construct advanced scientific instrumentation, our last century has been one of unprecedented expansion in our knowledge about the whole range of the molecular world that is our environment; from geology to biology and out to the farthest reaches of cosmology, our educated minds have taken us far. We have learned the workings of planetary evolution in solar systems and cosmic evolution in galaxy formation, we have glimpsed the life cycle of the stars. It seemed as if the trajectory of the species was such that, all in all, we were heading to a good place with our dedication to technological progress and scientific pursuits. In spite of the chaos and violence that dogs our way through history, the overall arch of the story was such that it looked like it would all be worth it in the end.

The ecological crisis has changed all that, causing us to profoundly question this definition of progress that lead to so many cultural and industrial *cul-de-sacs*. Though the full ramifications are still rippling, eventually not a single aspect of modern life will be left unaffected by what is being wrought. It is not difficult to see how these changes question our society's central dogma of progress through colossal technological means. Our societies have justified their activities by appealing to this definition of progress. The concomitant economic idea that the rising tide was going to lift all boats has been crucial in justifying the extreme income disparity this type of machine-driven progress has made possible. Now that the rising tide is recognized as a tsunami strengthened by climate change, a crisis of meaning is spreading. It questions the fundamentals on which the global social order is constructed. People of goodwill and with careful intelligence are questioning the ultimate worth of fossil fueled industrialization with its ubiquitous, life harming infrastructure and attendant justifications in growth economics and consumerism. People are even questioning the value of capitalism and democracy now that they seem incapable of dealing seriously with the ecological crisis.

I am suggesting that it is vitally important to understand the psychological roots of the ecological problems facing our cultures. In our minds, at some level, we are toying with the idea of collective suicide as a way to stop the pain of existence. We see this most directly in the popularity of apocalyptic cults but its tendrils are everywhere. As we enlarge and refine the destructive capacity of our nuclear arsenal we assure ourselves that power rules and compassion is for fools. It is what the idolization of consumer selfishness has lead to: whole societies that live cut off from heart healing experiences in the natural world and the sense of safety and security that come from the balms of human kindness. Is there no escape? Those who are convinced that might makes right insist that this society is wholly incapable of changing its giant sized industrialized

What Is Mindful Ecology?

ways, those ways that grant the few winners of our neoliberalism their current intemperate privileges. Under this iron doctrine the value of all blood is measured in dollars and cents. Our instincts as human beings know this dismissal of individual dignity is not right, and that it can only lead into dark and dismal futures, but we don't know how to stop it without hurting what little sanity and stability remain. We are like a caged animal desperately chewing off its trapped leg to break free before the hunter returns to deliver the killing blow.

In some ways the modern world is a victim of its own success. It is our technology that allowed the human population numbers to explode exponentially while simultaneously increasing the size of the ecological footprint of each individual in the overdeveloped world beyond all reason. Our technology provides the health care that keeps more people living longer than at any previous time in the long history of our species. Due to our engineered and social environments modern knowledge have made possible, for the vast majority of human beings alive today these are higher quality lives than could be expected otherwise. Literacy, education, and opportunity have pervaded the globe as the percentage of people living in absolutely desperate poverty continues to shrink due to trade and globalization.[1] This is quite a meaningful and very real accomplishment, one for which we as members of the human race should rightly take great pride. The hope in our ability to transform the best of what we have accomplished is what inspires Mindful Ecology. A reorientation is needed because, unfortunately, this globalized development as we have practiced it so far has been accompanied by a number of harmful effects as well. The obvious ones are the pollution of the land, sea and air. This is evidence that the great wild sinks of nature are no longer capable of composting our industrial wastes. Equally worrisome, if not more so, are the less obvious harms that have developed over the last centuries: those that are found in our communities and in our own psychology. We are at war with the planet's animals, our biological kin, and at war with the very water and air and soil we use to weave our nest. We are at war with ourselves.

To find a cheerful heart, if we are not to succumb to the Pollyanna but keep our eyes open to the facts, will not be an easy path to travel; not for individuals, nor for our communities. The innocence of our Western inheritance has been lost for those with the courage to look upon the end of its current chapter. Further movement along the same well-worn ruts of business as usual will not turn out well. Regardless of what we collectively choose to do, or not do, the lifestyle of the industrialized world cannot physically be maintained much further into this century; there simply is not enough oil, rare earth metals, potable water, and soil remaining, nor enough social capital to avoid the resource wars this entails.

So what are we to do?

My Story

People are being emotionally torn apart by the ecological crisis. The desperation in this state of mind is powerful and needs equally powerful tools. The ancient practice of contemplation provides one of them. I do not believe the society we are living in is sustainable, so I see no use in "fixing" people who are deeply troubled by the horrors of the crisis so that they can fit back into it better. I have chosen to walk further into the awareness of ecology to see where it might lead. This is what is being offered here, a way to listen to the earth herself. In this I trust. In this raw, direct encounter between the earth and people of goodwill, I see a joyful way forward. Not one without pain, we are far too along for that, but one where the pain is no longer allowed to crush the colors out of the untouchable purity of deathless life ceaselessly pouring forth all around us.

I entered years of despair after studying piles of books about peak oil, species extinction, limits to growth, system science, soil loss, ocean acidification, over-population's over-shoot and all the rest of it. Outwardly I kept on for the sake of those whom I love and who depend on me. To keep a good countenance, the basic cheerfulness every good home needs, was at times a serious challenge. I had grown up meditating and kept to a regular practice of some type until my marriage. Though decades were spent playfully raising our children, concern for their future was never far from our thoughts. When I entered my ecological study period it was inspired by my deepest roots and commitments. It drove me back to a regular meditation practice. Eventually the hurt which I lived with day in and day out grew to unbearable proportions. I began looking around for something more I could do when I had the opportunity to attend one of Joanna Macy's workshops known as the Work that Reconnects. That was the turning point. By expressing the grief I was able to rebuild that which had been broken.

In my experience no one can tell anyone else what they should do or need to do for the earth in these desperate times. What we can do is work to make our aspirations as clear as we can, admitting them to ourselves in all their hope and fear, and then nurturing our simple love of this earth-life experience with all its ups and downs each and every day. Nurture biophillia, help it grow within your mind-stream. Do this and I trust the rest of what needs to happen will fall into place for you too.

It is my working hypothesis that the more people who do this attunement with the earth, the less painful the transition out of our unsustainable culture will need to be. It is also my conviction that doing this inner work of taming the mind is a very direct way of working on the types of skills that will be ever more in demand as the limits to growth continue to bite and suffering increases. The times are late and the needs to relieve suffering are at hand all around us. Clear your head of cobwebs and make-believe, walk away from the dying Empire and wake up to the preciousness of the moment.

To try and capture the position I arrived at in somewhat poetic form I might say something along these lines:

If we had to place our final confidence in the mass of mankind, well, that would be a rather dismal prospect. Mankind itself is in the hands of something much greater. We are undeniably interdependently related to the whole of the earth's biosphere. This is a scientific fact, not a theological or metaphysical assertion - we are in the hands of something greater than ourselves. In Mindful Ecology we apprentice in the art of placing our confidence in that greater. It does not encourage subservience, let alone groveling before the immensity and complexity of a living universe. It encourages respect for the nobility of one's own existence, the most intimate and immediate example of the true nature of the universe we will ever know. In our own subjectivity we find equality with all sentient beings. In that equality we are liberated: the immensity of father's stars no longer frightens us with vertigo and the complexity of mother's organic life forms no longer overwhelms and scatters us. We learn to take our seat, here, in the middle way. We seek to become ever more authentic human beings. In carving our own name into the pool of earth's many causes and effects, we find we have had a home prepared for us "from the beginning." Wisdom joins hands with compassion as slowly, somewhere far from the reach of changing circumstance, a small smile forms, for in our hearts we have cleared our shrine and placed on it our offerings of "yes" and "thank you."

This teaching is medicine. What is being offered here is a method by which open-hearted people torn apart by the increasingly bad news in our ecology headlines can avoid suicide and despair. At least it did so for me; ecology, I found, was a path of sorts. Carl Jung in *Modern Man in Search of a Soul* was convinced that dehumanized mass industrialization had brought about a type of mental illness which he referred to as a loss of soul. This is an apt account of what I think is affecting many of us today who find our spirits crushed and our hearts broken by what we have learned from our studies of the earth sciences. This is what learning about the wall-to-wall suffering our way of life inflicts on all the rest of life eventually did to me; I found I had stepped through a very dark threshold. Working in the corporate shells of America in this age of runaway financial shenanigans only exasperated my sense of nihilism. Things were growing acute when the important revaluation struck. I realized that most

What Is Mindful Ecology?

fundamentally I wanted to turn right side up what seemed, to me, to be upside down. I was in this great depression because I loved this life on this beautiful earth so very much that it pained me to see how what is most precious to me is harmed and abused, poisoned and broken. Meeting people like Patch Adams, JoAnna Macy, and my Buddhist teachers and friends, I began to learn how to let that consciousness that loves the earth so clearly be the ground which the exercise of my intelligence and will acknowledges. It was a fairly small adjustment, but a difference that made a difference.

My time with my loved ones and my loved garden is short. I want to be free to be able to express my thank you for loving me, and letting me love you, by being happy. I could clearly see it would be better to try to be a happy companion, someone who brings whatever energy of light and joy I can muster to the day to day life we share. The critical voice had become the primary voice. I needed to change that, to get 'thank you' and 'yes' back into the primary role in my experience of life. Learning to think slowly and carefully about things in a way that included my whole physiological and emotional being, instead of just my conceptual mind, took place with a certain gravitas. By retraining my brain to start from a place of devoted gratitude to the earth for every day I get to play in her wonderful graces, my whole life slowly changed. I trust this same process will happen to anyone whose spirit has been similarly broken. This book is to encourage just that.

Buddhist Influenced Western Contemplation

Mindful Ecology uses meditative and contemplative practice to engage with the ecological crisis of our age on a personal level. Using Mindful Ecology people are doing more than just reading or hearing about the ecological crisis; they are fashioning their hearts and minds around it. This practice naturally encourages us to extend our compassion to family and friends and, ultimately, extend it to embrace the whole of this evolutionary deep time mystery we find ourselves a part of. This is, after all, our most intimate nature, this DNA dancer dancing its extended phenotype as our earthly days play out under the vast starry sky.

Most scientists involved in researching the accelerating breakdown of planetary ecosystems are much, much more concerned about the consequences than are our political, religious and intellectual leaders. The trends in the data indicate a radical makeover of just about every aspect of life that people in the developed world consider a necessary part of business as usual. What is often referred to as consumerism is passing into the history books. These changes will be guided by a truism: that which cannot be sustained will not be. As a society we are not at all well prepared for dealing with changes of this magnitude. This leaves the caring, educated individual as the focus for whatever alleviation of the suffering such large changes entail.

Mindful Ecology looks for means to strengthen our personal engagement with the earth as we experience it. This is an important point. Mindful Ecology is grief work; it seeks to increase our ability to be with the truth of our time which is dark and filled with pain. But it does not stop there, this is not at all the whole story. It uses this grief as fuel for an alchemical transformation of both our body and our mind. The path is a means by which each of us can find our own way to a relationship with the living earth, the gateway to which is our own bodies and their internal and external senses. This strengthened engagement with the earth not only sustains us through the dark night, but also brings joy and happiness to our day. It is a wonderful thing to be here together, sharing in the reality of the here and now. It will not always be possible.

It is a sad fact of human psychology that we often do not appreciate something precious to us until it is gone. Mindful Ecology is about waking each other up now, before we lose even more of the biological diversity and wholesome ecological systems on our planet. When we wake each other up through prose or ritual, poetry or contemplation, it is to recognize with heart-aching clarity the precious-

What Is Mindful Ecology?

ness of life on earth, a preciousness we should not take for granted as callously as we do. I do not subscribe to the happy chapter at the end of a typical ecology crisis book. I am from the Dark Mountain tradition. I believe this civilization built on the cheap energy of fossil fuel is collapsing in a process that will continue for centuries. The way down will, however, continue to be punctuated by dramatic breakdowns of overly complex infrastructures and institutions, such as those that fill our headlines already. It is not going to be the end of the world; it is the end of what is known among some ecologists as Homo Colossus: giant sized fossil fueled infrastructure and industrialization.

It is not possible to discuss these issues without acknowledging that they share many roots with religious concerns. The icons and mythic stories of our ancestors deserve respect, even when we hold them accountable for institutionally encouraged crimes committed in their names. There has always been a meaningful difference between those who use our deepest symbols to enslave others and those who find liberation under the tutelage of those same symbols, if I may speak so. This is the case in both Eastern and Western religious traditions. Many who took the 1960s Journey to the East woke up late in life to a new appreciation of what the karma teachings had been all about. It is a mistake to think we can simply disregard the traditions which saturated the cultures we grew up in. We absorb the tales with our mother's milk and they too enter our bones. That said, modern awareness is global. Mind, curious and sensing the danger up ahead, is exploring for truth throughout the many rich and diverse cultural heritages our planet hosts. No longer do the provincial choices of a small town limit our spirit's desire to share in the sacred as our brothers and sisters from the next valley have understood them.

Mindful Ecology, though rooted in Western thinking, looks also to Buddhism for inspiration. In Buddhism it finds both a path of mind cultivation and an appreciation for the happiness a simple lifestyle allows. Eastern Orthodox traditions of Christianity hold many more teachings about silence and breath but the Western brand of goofy

700-Club-influenced Christianity leaves most people wholly in the dark about contemplative practices. (Learning from our brothers and sisters from the East might also help us lose that chip of our shoulder consumerism wedded to triumphant Empire building has bestowed on so many people in the West in spite of good intentions.) Scientifically, the only viable response to the ecological crisis is one in which people learn to build good lives using less material, energy and distracting stimulation. These goals are embraced by what has been called Buddhist Economics[2], which is the study of economics on the human scale. It can provide a realistic framework in which individuals can work out their own way of dealing skillfully with the most important challenges of our times. At the heart of this tradition is the idea that we should all lead meaningful lives by employing our time and energy in ways that contribute to the healing and well-being of all sentient beings. Buddhist Economics puts people first by valuing meaningful work well done above efficiency, cooperation above competition, and personal respect above the use and abuse of one another. It is my experience and conviction that people are happier when they adopt a more ecologically sound lifestyle. People want to be producers and not just consumers, only parasites do that. Grow a little food in a garden, learn to store a bit of it, and you might find it slowly transforms your life. There was a strong tradition of productive households in our lands before social engineering was used to remake the citizenry over in the image of consumers. It is time to revive the technology and sociology of the simpler past, in combination with the best of what we have and know today.

These Buddhist roots also protect us from straying too far afield into the prepper mindset where bullets and bread for one's own becomes the be all and end all of practical response to the fears of social collapse. The Rambo fantasy of hyper-violent revolution is just that, a fantasy. The zombie apocalypse has been canceled.

There is another reason to honor Buddhism in these reflections. Ecology is a science of interdependence, it is inescapable since within ecosystems we find that everything is connected to everything else. Buddhism has been teaching the interdependence of the natural world since its inception.

Finally, and most practically, it is in Buddhism that a long and honorable tradition of mental cultivation through disciplined contemplative practice is found. Depending on your own path it can be important to have someone who has been around the block, learned a few things and knows the score to teach you how to properly begin a meditative and contemplative practice. As mentioned I believe we in the West have much to learn from the East about the power and value of cultivated subjectivity. We in consumer societies have much to learn from the contemplative monks and nuns, East and West, about contentment and the secrets of a happy human life.

Mindful Ecology also draws inspiration from the religious traditions of the West. There is no denying the cultural inheritance by which the modern world has come to be what it is. There are many more treasures to seek out around creation spirituality within the Judaic, Christian and Muslim faiths than have heretofore been acknowledged. The apocalyptic theme these religions share also plays a role in our modern societies that is altogether too central to the concerns of Mindful Ecology to ignore. I agree with Carl Jung who noted that it is, "Only by standing firm on our own soil can we assimilate the spirit of the East."[3] No lasting psychological health or maturity will come from our Journey to the East if all we do is paint ourselves in the colors we find there. It might seem what the monotheisms have to teach the world and what the non-theistic Buddhist philosophy offers are forever irreconcilable. Fruitful dialog might seem impossible between those that teach reliance on God and those who insist that being a fully compassionate human being is sufficient. We will see. In my mind these things are not as irreconcilable as they might first appear if we are willing to allow science and psychology to also come join us around the table in our conversations.

It is one of the convictions of Mindful Ecology that our troubles with ecology are related to a misreading of our own inherited mythologies. We have become scientifically astute but symbolically illiterate. Surrounded by the greatest debasement of imagery in the history of the human psyche, thanks to the pervasive penetration of the mass media pushing the lowest common denominator, the minds of most people, most days, is not a particularly happy place to be. Not happy inside our own skin, we are not as worried about losing it as we should be. Due to interdependence, the outermost layer of our skin is the biosphere itself. This blue jewel against the night sky, which is in fact "our ancestral home of old,"[4] is being treated as a consumer good designed with planned obsolescence in mind. It is not. There will be no replacement if we continue to treat it as a product destined for the landfill.

Our social contracts are being tested. Racism and religious fundamentalism are on the rise. Governments and the corporations they serve are becoming increasingly aggressive in the pursuit of their power and gains. It is as if we have a bully in our midst and it will keep pushing us to see how far it can until we stand up and say, "Enough!" It is a conviction of Mindful Ecology that there is no hope of achieving a lasting peace with our circumstances or with each other until we are able to talk about what it is that is actually ailing us. War will cut our numbers but solve nothing in the long run. The planet will still continue to warm due to our tailpipes and the oceans will continue to die due to pollution sewn into the biosphere decades ago. Mindful Ecology is asking if we are ready to have an adult conversation about the full implication of these things. Yeast will overshoot their numbers if they are given the resources which provide them the chance to do so. Human beings will, evidently, do the same thing. Can we learn to live with that? Can we learn to celebrate our close familial relations with the rest of the fruits of DNA and deep time, or will we insist we are too special and only a grand apocalypse is worthy of our stature?

The very stories that were meant to encourage us with the message that in the end the good guys win and that the path of compassion is wisdom itself, both the Buddha and the Christ stories, might be twisted into the vehicle by which the bad guys are empowered to do their worst. The central conviction of Mindful Ecology is that the

What Is Mindful Ecology?

wisdom traditions of the East, West and Indigenous peoples are in one accord in teaching that heartfelt gratitude for our earth is the proper response to the gift of existence and the cultivation of virtue the proper path towards an honorable and noble human life. These traditions are means of guiding the mind, which is so easily mislead and gullible, to an encounter with reason and reality. This is where love is found. This is the birthright of every human being.

The Ecological Crisis is a Symptom

Mindful
To be mindful is to be aware of what is happening right here and right now.

Ecology
Right here and right now human activity is threatening the future of our species.

Mindful Ecology
To be aware, right here and right now, of this ecological truth.

If one is convinced that the life we live now is unsustainable and has no future, what is a person going to do with such knowledge? It seems that there is so little an individual can do in the face of our collective choices driving our society to make a bad situation worse. If the only hope is for our society to wake up and start making sense, well, that I fear is not much of a hope at all. It looks all the world like our societies are insisting on pressing this un-sustainability just as far as it can go before crying uncle. Where then should we look for real work that just might be of real benefit to ourselves and, equally importantly, might really benefit the next couple of generations that are going to have to live through environmental hell? What can we do? As it happens, I believe we can do the most important thing of all. I believe it is only in the world of individual lives that the true balm for what ails us is to be found.

I am presenting the argument that the ecological crisis is a symptom. The disease is in how modern ideas about mankind's role in the universe have poisoned our relationships with each other and the rest of the living and non-living world. My position is that we do not know our own minds well. We do not know what they are capable of in peak moments of bliss and cognitive clarity, nor do we comprehend how easily they can carry us away on delusional abstractions that have no basis in the reality of our molecular world. We are so in our heads we are at risk of losing touch with our body, the physical reality in which our lives unfold. We are so into our abstractions of economy and nationalism, status and hierarchy, that we are losing touch with our need for clean air, water and soil. To be risking even the slightest chance of the kind of planetary chaos ecologists are warning us about, with the calm demeanor of our existing social discussions is, in my mind, a sure sign of collective psychosis.

Before we will be able to bring our collective engineering creativity to bear on the real issues we will need to be able to speak about what they actually are. Warming up the nuclear weapons to grab the biggest pieces of the shrinking pie is the alternative. Though sadly it is the alternative chosen by those currently in power, this conversation is not over. Those of us who still place our hope in reason will never be bullied into bleating; we insist the argument around the ecocide claim continues. Due to the burden of evidence behind the claims in the ecological, biological and evolutionary sciences, the burden of rejoinder is now on their side in the court of public opinion. Though our leaders do not seem to understand this yet, it would be premature to assume they never will. Facts, as mentioned, are stubborn things.

What Is Mindful Ecology?

Mindful

Traditions of meditation and waking up are surrounded by a wall of Cosmic Foo Foo so high it is almost impossible to see the point. I am suggesting that most of that is just not relevant to what these teachings are really trying to convey. What they are trying to say has everything to do with waking up to the reality of your situation as it is in the here and now and as it will potentially, yet realistically, become in the future. We have a funny habit of making ourselves slaves to our own ideas about ourselves and our world by forgetting that as long as we draw breath we are free to remake today in any image we desire. People walk away from lifetimes spent in abusive marriages, totalitarian religions, and hateful ideologies all the time. One day some insight dawns and they see through some ignorance on their part which had lead them to believe in their own slavery. At that moment they are free and there is no going back because the truth of the insight is always right there in front of them, reflected in the ceaseless unfolding of reality as reality. There is no going back into the cocoon of make-believe.

I am convinced that the same one-way insight comes into the lives of everyone who plumbs the depths of what our ecologists are saying. This ecological insight is already pervasive, like a shadow running through our modern world, and it is spreading. More and more people, young and old, will be caught and forced (or is that called?) to plumb these dark and depressing depths in their search for the truth. Mindful Ecology recognizes this process as a full blown hero's descent into the underworld, the first step on the road to enlightenment. It offers encouragement for a serious, daily practice of meditation and suggests finding a like-minded community. It is by developing the skills of meditation that we are able to begin to integrate our head's knowledge with our heart's responses to that knowledge. It is my position that we are in crisis due to a disease within our minds and bodies that cuts us off from our ongoing experience of the living earth. It is a disease with a long pedigree in the non-indigenous cultures of the East and West but one that can be cured within the individual. Mindful ecology seeks to live once again in a sacred world as our ancestors once did. That is what the world looks like to one who is awake.

Ecology

There are a number of lines of evidence that have led me to my position about the high likelihood of a collapse of the existing global political and economic order. Most of those lines of evidence deal with the relationship between human society and the natural world, that is, they are drawn from the science of ecology.

In a subject as large as this it is not possible to arrive at detailed, definitive conclusions. The best we can hope for is that by applying careful thinking we are able to provide reasons for or against our assertions. Those reasons can then be examined and each person can make up their own mind about whether they agree with the reasoning from the evidence or not. What is not allowed is to dismiss the claim that we have entered the age of ecological blowback by simply refusing to examine the evidence, or insisting that "we will think of something" when confronted with news about hard limits to the growth of industrialized society.

The fate of the earth just might hang on the results of the debate between economists and ecologists.

The popular story about our ecological condition is that it will be possible to convert our existing infrastructure, including our most enormous industrial machinery, to alternate energy sources. It must be so, we tell ourselves, since there is no other way to continue our existing social arrangements, particularly the globalized corporate production of consumer goods and the endless growth economic model required for fiat currencies to function. Those who claim there is no future in which consumerism can remain the justification for our activity and the organizing factor behind our social and economic arrangements are dismissed as extremists.

What Is Mindful Ecology?

The less popular story about our ecological condition is that it will
not be possible to convert our existing infrastructure to the sustainable use of energy and material. By this way of thinking the modern
industrialized world has already over-reached what the planet can
provide and is only functioning still by causing irreversible damage
to vital ecosystems. Though this characterization of the ecological
position may sound extreme, it is none-the-less true in so far as
there are numerous characteristics of the ecological crisis described
with detailed documentation in the scientific literature that will
seriously threaten the ability of globalized society's existing institutions and infrastructures to continue functioning over the next few
decades - decades - not centuries.

Mindful Ecology

I think our societies understand the message of our ecological
sciences quite well but are completely lacking in a proper public
response. Mindful Ecology attempts to address that gap. It willingly
struggles with the emotional aspects of the ongoing ecological crisis
and how it saps the meaning out of existing human activity. After
all, if our current hustle and bustle is only accelerating a drive over
a cliff it is hard to want to continue participating in the madness. At
just this point the individual is thrown back on themselves and they
must decide if their lives will be changed to reflect their new understanding. That is where the rubber hits the road: do you think this
crisis is important enough to actually change the way you live?

How might we express just how important this debate between
the ecologists and the economists might turn out to be? It is not an
academic debate. There is a cost in human lives. A lack of potable
water is threatening to kill millions, if not billions, of human beings
over the coming decades. There is a cost in lives disrupted. The mass
migration of refugees from resource wars and lands made inhospitable to farming are already filling our headlines. In the coming
decades these numbers are likely to include millions more as whole
landmasses become inhospitable due to drought, desertification

and rising sea level. There is a cost in wisdom and treasure lost. People desperate to meet the basic needs of food, water and shelter are not at liberty to continue cultivating the history of ideas or maintaining the complex arts of engineering and craftsmanship. There is a cost in social anomie. Though communities often pull together when disaster strikes, long term hard times do not tend to bring out the best in people. In a hot, over-populated world of shrinking critical supplies we can expect war, xenophobia and fundamentalist driven violence to rule the day.

Civilizations have collapsed in the past, we know this and have written grand sweeping histories of the world in which their rise and fall are plotted out in great detail. So far planet earth has seen the flowering and fall of twenty three civilizations. Knowing this, how can we not help but worry about ours succumbing to the same fate? This count of civilizations comes from historian Arnold Toynbee whose epic multi-volume A Study of History celebrates and records how each civilization expressed its own unique genius as it rose to some peak of flowering culture and then fell, collapsing under the weight of complexity and, as we understand so well today, under the burden of their unsustainable ecological exploitations.

The question each thinking and caring person needs to grapple with is whether or not our civilization is sustainable. If we find it is not, what could be more important than experimenting with our own lives, asking ourselves how then should we live? Mindful Ecology hopes to contribute to that spirit of experimentation.

Doesn't the answer to that question also change the whole approach one makes towards working on ecological issues? After all, it is one thing to pursue green technologies within the on-going project of spreading consumerism over every part of the globe, and it is another thing altogether to admit that there is nothing that can be done to avoid the disruptive changes heading our way and so our wisest course of action is to learn how to hunker down and adapt to them. Mindful Ecology hopes to offer tools for those involved in the triage tents surrounding this war on life; for the sixth mass extinction continues, the oceans are already dying, and the climate change that was prepared for us decades ago is already spelling the end of business as usual. God help those who will live through the climate change we are preparing for them.

It is time to wake up.

This is Mindful Ecology

Mindful Ecology accepts the reality of this molecular world as it is understood by science and trusts careful evidence based reasoning. It suggests a model for what ails us and a cure. The ailment is cognitive distraction and emotional exhaustion brought about by saturation in images of mass communication. Ever changing headlines, all delivered with the same tone of urgency, make hash of the ability of the mind to differentiate what is meaningful and important from what is not. So the path is to learn to quiet the distracted mind so that it can learn to think carefully again. Contemplation in a still body is a process of thinking we can call full-bodied reasoning. Contemplatives put the findings of neuroscience to use by recognizing that proper reasoning necessarily involves our emotions as well as our cognitions.

To put it rather glibly, we are confused because we are not as clear as we think we are about the difference between the jingle in our head that assures us 'Subaru is love' and the factoid in our head that teaches us the carbon content of the atmosphere is such that climate stability is threatened.

First the mind teaches the heart, as what we learn becomes felt in our body. Then the heart teaches the mind, as what we know in our bodies begins to inform our minds. Then we hear, each in our own way, what the lands, skies, and oceans are asking of us.

2.
Earth's Ecology

"In a future that is as unavoidable as it will be unwelcome, survival and sanity may depend upon our ability to cherish rather than to disparage the concept of human dignity. My purpose in writing this book has been to enhance that ability by providing a clear understanding of the ecological context of human life."

Overshoot: The Ecological Basis of Revolutionary Change
William Catton

Overshoot

Mindful Ecology is primarily concerned with the subjective side of the ecological crisis. It assumes that a study of ecological issues from reputable scientists and reporters is an ongoing activity for those who are drawn to care deeply about the state of the biosphere. This is a wide and diverse field and there are any number of extremely important points that could be made about what has been found. Surveying my pile of books, and considering what has remained central to what I have learned over the decades I have been involved in studying these subjects, one author stands out for me as providing an essential model for grasping what the core facts are teaching us. William Catton's *Overshoot*, first published in 1982, rather succinctly cuts to the chase. Allow me to introduce you to his model of Homo Colossus and the magic of phantom acreage that sets up a population overshoot followed by collapse.

People are lulled by 'oh, they will think of something' and 'it's not that bad' and 'we still have plenty of time' and bushel-loads of other thought stoppers. Professor Catton's work can correct all that. Understanding how detritus feeding leads to overshoot, how oil's unique energy density fueled the build out of the life threatening industrial giantism he christened Homo Colossus, and how phantom acreage has functioned to place us up a creek without a paddle, already far beyond a tipping point - all this is the simplified scientific message of ecology as I read it. It sidesteps the greenwash and states it like it is. These models and metaphors from Overshoot capture what is essential.

Most people today, obviously, have not studied the issues around ecological problems from the detailed perspective of an engineer being asked to "fix" them. Nor are most people going to be able to master all but a small part of the mountain of observational and theoretical information available today that describes the ecological conditions throughout the planet. This makes it difficult to see the big picture. This is where a handful of key models can provide a comprehensible structure. A few simple mental models, carefully chosen, can provide us with an indispensable common language by which we can address these things both with others, but also within our own thoughts as we try to make sense of what we know and the events of our times.

The ecological models we are about to touch on are among those that provide a basic ecological literacy. The material in this chapter is intentionally dense, it is meant to provide a shared vocabulary. For the full evidence and argument there is no substitute for reading Catton's masterpiece and ecology's classic textbooks.

Thermodynamics

An ecosystem is a name for an organized unit, a logical level that is complete in that it includes all the components it needs to survive over the long term, however that may be defined for different purposes. Ecosystem models are created when a boundary is drawn around the functions of interest; a patch of garden, pond, forest or planet. As soon as boundaries are introduced a system is defined. Ecosystems are open systems, which mean these models explicitly include interactions with their environment. There will be inputs, typically energy and raw materials, and outputs which typically include thermodynamic waste heat and processed materials.

Since ecosystems model the earth's biosphere it helps to have a clear internal reference of our planet's position in both time and space. Let us remind ourselves to include what we know about the earth's temporal and spatial environments when bringing it before the mind's eye, for everything has evolved together, including us.

In the early solar system orbiting dust grains collided and stuck together in a process of accretion that in approximately 10,000 years produced boulders and asteroids a kilometer wide. Over the next million years these objects continued to collide forming moon and mars sized objects. These baby planets crash into one another over tens of millions of years until there were just a few survivors, each in its own orbit. So far this is all standard stuff from a high school astronomy class but to begin to pierce the mist of time and absorb your ancestry in your bones it might help to contemplate two details of the process, seeing them as they might have unfolded; the formation of our moon and the arrival of water.

When the rocky inner planets form, the denser elements sink into their planetary cores. These iron and nickel cores support the less dense molten magma consisting of rocks rich in oxygen, silicon and such. Above the magma the planetary crust forms. Some 50 million years after the accretion began early earth collides with another baby planet with such titanic force that it melts the crust and sends vaporized rock orbiting our young planet. In this final major accretion event our moon was born. The vaporized rock collides and sticks with itself until our companion is formed, roughly 25% as big as earth but huge on the horizon with an orbit only 10,000 miles away. The moon has been receding from the earth ever since. The moon, uniquely in our solar system, lacks an iron core since by the time its birth collision occurred these heavier elements had already sunk to the earth's core. Only the magma rocks were ejected.

Earth's Ecology

As the great gas giants of the outer solar system complete their formation they perturb the orbits of the meteors and asteroids. On earth the bombardment becomes extreme yet it also brings water, the essential element for life. Only objects far enough from the sun are able to contain water that is not boiled off, far enough away to form ice. Out between Mars and Jupiter today we can see one of these asteroids, 1-Ceres. At close to 1,000 kilometers across it is nearly round, a proper planetoid, but not very dense probably because it contains a large amount of water ice. The earth's waters, covering 70% of the planet, could all have arrived here in collisions with just a few such asteroids. During contemplation picture in your mind's eye the arrival of these bubbles of life giving water on our fiery, volcanic planet until a natural awe and gratitude arise. It is difficult to pierce the mist of time but we have a knowing, a type of intuition about what we are. We can sense the long chain of cause and effect from which we come. *Our ancestors were titans.*

What is arguably the most famous photo of all time was taken December 7th, 1972 by the last manned mission to the moon, Apollo 17. It captures the full diameter of the earth as seen from the moon, the blue and white globe so majestic against the darkness of space. See it in your mind's eye, see how absolutely self-contained our planet is materially. The mass of our planet was gathered approximately 4.5 billion years ago, and aside from a few meteors here and there has not substantially changed its material content since. All life has ever had in the past or will ever have in the future to survive and thrive is here on the planet right now. Ecology teaches that materials cycle. They are used over and over again without losing their ability to function. It is the ongoing miracle of the molecular world. All materials have their circular paths, like water as it moves from ocean to cloud to rain to river to ocean to cloud. . .

Materials cycle, energy does not. Energy is a one way flow which can be temporarily captured, diverted, used to build complexity and sustain life as anti-thermal dynamics yet inevitably, in total, will always drive towards an increase in entropy, towards a more dispersed, useless state. Energy cannot be reused. It can be transformed from one form to another, as we see most astonishingly when photosynthesis converts light into food, but every transformation will only proceed if there is, on the whole, a degradation from concentrated energy to a form more dispersed and dissipated.

Earth, our jewel in space, is continually bathed in the light of our sun, bathed by radiation about 10 percent ultraviolet, 45 percent visible and 45 percent infrared. This unceasing flow of energy provides the one way gradient on which the web of life weaves its majestic forms.

All the ecosystems on our planet depend on the energy received from the sun, aside from a few specialized ecosystems such as those that use the energy of geothermal vents. These ecosystems structurally consist of the primary producers and the secondary consumers, the plants and the animals. The primary layer is able to fix sunlight for the manufacturing of food from inorganic materials; green plants, algae and water plants. This biotic component is called autotrophic, which means self-nourishing. The secondary layer is heterotrophic meaning other nourishing. Since heterotrophs are unable to create their own food they must acquire it by consuming the complex materials created by the autotrophs. This single fact contains much that we need to respect to lead a noble human life.

The secondary, consumer layer is usefully further divided into herbivores, carnivores, omnivores and saprovores. The herbivores eat only plants, carnivores feed on other animals, omnivores feed on both plants and animals and saprovores feed on decaying organic materials, detritus. Most people have encountered these terms before, except perhaps the term saprovores which is a touch ironic considering petroleum is decaying organic material. Saprovores feed on decaying organic materials. When humanity started its dependency on non-renewable fossil fuel energy sources it entered into a detritus ecosystem. These ecosystems are characterized by exuberant population growth followed by a population crash, a die off. More about this as we proceed.

The recognition of autotrophs and heterotrophs provides more than just a classification scheme. By following energy relationships through food webs it also uncovers the fundamental structure of earth's ecosystems.

The primary trophic layer of green plants supports the herbivore layer which is known as the primary consumers (the mouse eats the grain). The carnivores that eat the primary consumers are known as the secondary consumers (the fox eats the mouse) and finally, in some ecosystems, there are tertiary consumers dining on the secondary consumers (as the eagle scoops up the fox). Each layer is able to utilize only about 10% of the energy transferred to it; about 10% of the energy is converted into biomass. This creates what is known as the energy pyramid with a large base of primary producers supporting increasingly smaller layers above it. For example, in a simplified model a patch of field with 1,000 grams of wheat could support 100 mice as primary consumers. In the field 10 foxes could survive as secondary consumers on that many mice and those foxes could support 1 eagle as a tertiary consumer. All terrestrial and aquatic ecosystems are structured in this energy pyramid form.

Another way to track energy through an ecosystem is to look at the respiration rate in relation to the total production of biomass. Any complex structure above absolute zero temperature requires, as Schrodinger has shown, a continual pumping out of the disorder to maintain its order. This process, negentropy, negative entropy or what we today call information, has been proposed as the very essence of life itself.[5] In ecosystems, the complex biomass structure is maintained by the total community respiration which, we could say, pumps out the disorder. The ratio of total community respiration to total community biomass (R/B) is the maintenance to structure ratio, the thermodynamic ordering function. Nature might be maximizing this ratio through ecological succession.

The important take away from this is that there are finite quantities of material and a fixed flux of radiant energy available on earth. All of our planetary processes are defined by these limitations. The thermodynamic energy laws give earth its characteristic dynamics, strictly delimiting what is and what is not possible. With these tools on our cognitive tool belt we are now in a position to begin to appreciate the concept of an environment's carrying capacity. But first we will indulge in a philosophical aside.

Compost Heap Speculations

It is worth pausing with the saprovores a moment to emphasize the role of the compost heap in the larger scheme of things. We live in a death denying culture, worshiping youthful beauty above our elders' wisdom. We prefer our death delivered with all the make-up, staging and lighting an action movie can deliver, all wiz-bang and drenched in nonchalant violence. When scary death does show up in our horror films and books, it is typically draped in occultism's supernatural concepts around religion's hellish fears of eternal torture. We abide the zombies, vampires and aliens because we know about the real monster, the real fear we are forced to face, the one we all understand. We rightly fear, given the evidence of the last century, that instead of meeting what Robert J. Lifton in his study of Hiroshima and Nagasaki survivors called "regular old death," there is an increasing probability that one's life might meet with a premature end as a result of industrialized cruelty; be it the concentration camp, cluster bomb, a climate change disaster or thermonuclear war. These fears are legitimately horrifying.

Given our psychological landscape of images and metaphors for dealing with death and dying, it is not surprising that we draw a line around the act itself, removing it from our homes and placing it under the care of physician-priests in our hospital-temples. It is quite possible that part of why, evidently, we moderns do not sufficiently cherish life enough to protect it from the ravages of giant industrialization is related to how we view death. We seem to view it as unfair. We view death as an imposition to those as uniquely entitled as ourselves. Or we see it as a specter of evil, something that should not be that incarnates in man's cruelty to man. Where is the ancient teaching that death comes as a friend, a final gift from that which brought you forth? Perhaps we would do well to remember our death walks with us today, each precious day we have, despite our troubles. Modern man is at war with existence. Rejecting the reality of subjectivity and the irreducible value of the individual, our dedication to giantism has left us defenseless, philosophically orphans, little more than paupers on the fields of nihilism. Instead of death crowning lives well lived into old age, a death we can more easily take in stride 'to make room for the young uns,' we seek to "conquer" nature. In this fight against existence, ecology suggests we fight against monsters our imaginations have created. It is sad we chose to make some of them real.

Complex biotic materials are formed by the ever cycling materials of our planet and the non-stop one way energy flux of our star. When they break down they do not "die" in any ultimate sense. There is only the compost heap, the recycling of every element in making way for new life to flourish and in its turn, decay. Metaphorically we can say that there is no place cut off from the rest of the whole of Gaia in which the damned are cast off. Christian orthodoxy insisted that the Christ came to free death from the devil and hell, to liberate humankind by teaching us there is no such second death for those who love. The ancients taught the same good news in their mysteries when they whispered in the ear 'the sun at midnight is ever the sun.' And ecology insists that the dark humus of the compost heap is the farthest reaches of the truly existing molecular world for any and all of us born of earth. Being mindful of ecology we are able to hear that the ancient preacher knew a thing or two about Wisdom and Folly:

> "Shed tears for the dead man, since he has left the light behind;
> shed tears for the fool, since he has left his wits behind;
> shed quieter tears for the dead, since he is at rest.

> *For the fool, life is sadder than death.*
> Mourning for the dead lasts seven days,

for the foolish and ungodly all the days of their lives."
Ecclesiasticus 22.11-13, *The Jerusalem Bible*, italics added

Carrying Capacity

Carrying capacity is the maximum population size of a species that the environment can sustain indefinitely. In population biology it is defined as the environment's maximal load. The concept is critical if we are to understand two features of our world today. For an environment to sustain a population indefinitely its material and energy needs must come from renewable sources and there cannot be significant damage to the organisms or their environment because negative impacts lower the carrying capacity. The second critical point is that the carrying capacity of an environment can change over time due to changing conditions. Some of the many variables that directly affect an environment's carrying capacity include changes in the availability of food and water, changes in the ability of the environment to process wastes, and changes in the availability of energy in a form that can be utilized. In today's world all of these variables are changing in ways that are shrinking our planet's carrying capacity. This is the larger, slower reality behind the ephemeral headlines.

An ecosystem consists of a population of species and the changing environment they find themselves in. The environment is changed in large part by the very presence of the populations dwelling there. This progression through a series of stages is called succession. Each stage is termed a sere. Imagine a bare field and watch it evolve over a century or two. The bare field becomes grassland with weeds and other pioneering species, which in turn create the conditions for shrubs of different kinds to begin growing here and there. As the years go on the shrub density increases and the first trees of a pine forest find they have the conditions they need to survive. If the pine forest in turn gives way to an oak and hickory forest and the succession ends there then that last stage, last sere, is referred to as the climax community.

If the only major influences on the ecosystem remain those that come from the living populations, nature's use of energy can potentially keep the climax community running for centuries, even millennia. The climax community is not a wholly static metric of energy flow to production. Instead there are pulses which have been found to keep the populations fit. These climax communities are characterized by having the strongest adaptability to shocks compared to any other possible configuration of living things and their environment. Life alters conditions such that they become more capable of hosting life: the Gaia hypothesis. Their resilience is due to the climax community containing the highest degree of species diversity of both plants and animals that this environment can indefinitely sustain. Energy has brought forth the maximization of use before being dissipated into entropy.

The climax community is a theoretical construct for an environment that has achieved a configuration that maximizes its carrying capacity indefinitely. It is 100% sustainable.

In the climax community the amount of standing biomass is maximized but interestingly, this is not always what we humans want. Agriculture is a good example of deliberately holding back the process of succession to an earlier stage so that there continues to be larger yields of crops that we can use as food. Growing wheat, for example, keeps the land deliberately in the grasses stage. Additionally, just as one characteristic of the climax community is that it has wide biodiversity, a characteristic of the pioneering, invasive community is that it is dominated by a monoculture. There is a tendency for a single plant to dominate exactly as we see on our farm fields with their acres of wheat and corn. It takes an enormous amount of energy to pause the process of succession on our farms. The petrochemical dependency of agribusiness, from tractor fuel to fertilizer, is one of the most important real issues we should be publicly discussing. It is also well known that the right pest can wipe out an entire monoculture crop. Lacking the diversity of the climax community these critical ecosystems are vulnerable to shocks.

Still, even the climax community is not permanent. The larger change of sere ceases only until something from outside the system intervenes. In the real world everything changes; sooner or later a shock will come that is larger than what the environment can adapt to and it will regress or change completely. These shocks come from changes in either the inputs, the incoming energy, food, and materials or the outputs such as we see in pollution where something produced in the system overwhelms the capacity of the larger environment to absorb it and break it down.

When considering the value of this theory it is worth asking how the process behaves in extremus. One of the other characteristics of the pioneer, invasive species is that because they lack the feedback loops that would limit their population growth to sustainable levels, their populations are prone to a phenomenon known as overshoot and die off. The (in)famous example is yeast introduced into a vat of wine. This simple ecosystem displays many of the most critical concepts necessary for understanding the current ecological predicament mankind finds itself in.

In a wine vat the yeast will continue consuming nutrients from the sugary grapes and reproducing as long as they can. The more yeast, the faster the sugars are consumed. It is party time in the wine vat, what ecologists refer to as the stage or age of exuberance. As we have learned, all ecosystems are defined by limits and this one is no different. There is a limited supply of the sugar nutrients and the expanding population is drawing down that resource by consuming it faster than it can be renewed. This ballooning population eventually exceeds the carrying capacity of the vat with no way to replace what is being lost. We say the population of the yeast is in overshoot. Because this stage is unsustainable, it will not be sustained. There will be a population crash until the resources can recover to a level adequate to sustain the diminished numbers. In the vat the pollution that the yeast produce, the alcohol and carbon dioxide (our fermentation), fill the environment until the yeast is no longer able to survive. In that vat there is a die off to the point of extinction. There was a balance point in the vat between the number of yeast, their waste products and the ability of the mash to recycle them, but this balance point was exceeded, over shot. By exceeding the carrying capacity the yeast harmed the environment's future carrying capacity, creating a downward spiral leading it, in this case, to zero. These words form the key take away: drawdown, overshoot, crash and die off.

This ecological analysis highlights the difference between eco-systems that are sustainable and those that are not. To sustain the existing carrying capacity indefinitely, the community popula-tions can only be using renewable resources and those only at a rate that allows them to be renewed. This is an important point. Oil, for instance, is technically a renewable resource since new oil is being formed today, but the process takes millions of years. For all practical purposes oil is a non-renewable resource so with every barrel that we burn we draw down the total that remains in a usable form. Using any non-renewable resource is going to have the same characteristics. These are not methods of enlarging human carrying capacity, only exceeding it. Are there ways then that truly enlarge carrying capacity?

Another method is available. It is called takeover and humans are masters at this as well. It is the art of taking over other primary and secondary production for our own use. Every acre taken over for human settlement and raising our plant and animal foods increases the earth's human carrying capacity at the expense of other spe-cies. This is simply the way it is, a necessary result of life unfolding on a finite planet. History is a tale full of examples of one kingdom usurping the lands and resources of another. In these cases, all oth-er things being equal, the addition to the carrying capacity can be a permanent one. For most of human history this has remained the case because the contextual environment in which human settle-ment occurred retained its integrity. The larger global environment was able to provide the inputs we required and process our outputs successfully. We were fairly successful at diverting a large percentage of the world's life support capacity from supporting other life forms, towards supporting ourselves.

Around 500 years ago something new came to be. It had taken all of pre-history and all of history up to the year 1500 for our species to multiply to 500 million - half a billion individuals scattered throughout the globe (you might want to take a moment to really appreciate that). Even so, before the sixteenth century the European lands were full and population pressures kept the balance between the number of people and the carrying capacity of their environments more or less in balance. Though famine or plague would eliminate large percentages of the population, in time those populations would recover.

Then the Europeans "discovered" the new world, a whole hemisphere rich in natural resources. The largest takeover in human history commenced and an Age of Exuberance was begun which we are still living in today; an Age of Exuberance that is now coming to a close. Some nations seem to get away with supporting a population much greater than what could be provided for from their own lands. They did so by using ghost acreage, a concept developed by George Borgstrom. Such nations were drawing upon invisible carrying capacity, which is capacity located elsewhere on the planet. Professor Catton calculates there were approximately 24 acres of Europe for each European at the time Columbus set off. Life in Europe was a precarious affair with never quite enough to go around. After claiming the resources of the "new world" this increased to an amazing 120 acres of land per European.[6]

It took another three centuries for the human population to double. The globe sustained one billion of us around the year 1800. There are ecologists who study these things who think that perhaps about one billion people is the balance point of indefinite global carrying capacity for the human species in pre-industrial ecosystems. However that may be, the Age of Exuberance became the new normal. From now on, year after year, growth was expected. The Renaissance and the scientific revolution, the war for independence, and the drafting of the American Constitution all took place against this background of rising expectations. The new abundance supported widespread literacy for the first time as leisure hours spread among the peoples. Technological improvements became the definition of progress and many people began to feel entitled to a "perpetually expansive life" as Catton had it. A belief in limitlessness began to become common currency in our cultures, displacing the hard earned wisdom of previous centuries that had learned to respect limits. With the Western Enlightenment came a humanism that dared to assert that it would be possible by man's own ingenuity to improve his fortunes, improve the 'state of nature' and improve his social and political relations. The dead weight of superstitions which had haunted our ancestors was thrown off right and left, but so too were the numerous traditions which had acted to safeguard ourselves from our own excesses. An Age of Exuberance indeed. It is not difficult to understand how, again as Catton had it, we have inherited the conviction that mankind is "largely exempt from nature's constraints."

The Age of Exuberance would have run its course, finding a new plateau that maximized the benefits of the New World takeover, but for another fortuitous discovery. James Young, a Scottish chemist, noticed petroleum seeping out of a coal mine in Alfreton, Derbyshire. In 1847 he managed to distill light thin oil that could be used in lamps along with thicker oil good for lubricating machinery. In 1848 he started a small business refining crude. Mankind began a draw down not from elsewhere but elsewhen. Now we were dealing with another kind of ghost acreage, what Catton would call phantom acreage. It was the combination of the New World resource exploitation and the fossil fuel based industrial age working together that created the 500+ years of the Age of Exuberance and Abundance which we have now learned to take as the normal state of human affairs. In that special time of converging circumstance, never to be repeated, a new species was born in our midst: Homo Colossus.

Homo Colossus

"Man does not live by detritus alone."
Overshoot, William Catton

One of the unique capabilities that set Homo Sapiens apart from all the rest of the animal kingdom is our mastery of tools. As a species we are able to expand the carrying capacity of our environments through the use of technology. This has been going on for a very long time.

The important take away from this analysis for our purposes is how historically technology has performed the role of increasing the carrying capacity of our environments. The total carrying capacity could have been said to have been the product of resources multiplied by technology. Too much of a good thing, victims of our success, today we find that the technology we are using is actually shrinking our carrying capacity. It is no longer a multiplying factor. Today the relationship seems to be one of division. Now, it seems that the total carrying capacity is equal to resources divided by technology. This is worth a moment of careful contemplation. Malthus was concerned with the problem of expanding human population butting its head against fixed limits when he noticed populations grow exponentially but resources geometrically. In the real world we are discovering the expanding human population is butting its head not against fixed limits, but against the shrinking limits of collapsing ecosystems.

Increased population combined with our technology has now grown so large it is as if a whole new species has evolved, one which is very much capable of altering the biosphere as a whole. This new species was christened *Homo Colossus* by William Catton, capturing the essence of the challenge ecology poses for modern industrial civilization. To understand the powerful metaphor requires that we learn to look at man's tool use from another perspective. Normally when we think of our use of tools we consider them as means of

Earth's Ecology

adapting the environment to our human needs; we plant a farm of crops to feed ourselves, we warm our houses to fend off the cold. It is equally valid to propose that human tool use adapts humans to diverse environments. Our tools are somewhat like prosthetic devices we add to our bodies; we don a coat and now survive in environments that were formally too cold, we strap on a plow and fertilizer spreaders and find we can grow crops where previously the soil was too poor.

Our tools act as prosthetic devices; the cup of a mining scoop acts as an extended hand. At some point these prosthetics crossed the line into gigantism, where their sheer size began affecting whole ecosystems – a mountain removed here, a river diverted there. I do not think it is at all easy for the human mind to truly appreciate the scale at which human aspirations are unfolding all across the earth, all day, every day. That mining scoop just mentioned is capable of lifting 325 tons of "overburden" with every bite it takes into the earth looking for coal. This is not just a multiplication of men with shovels, but a qualitatively different event altogether. Consider the giant dump truck used in these mining operations. It is capable of hauling 380 tons of earth in a single load. The trucks weigh 1,375,000 pounds, rolling on tires that are roughly 13 feet tall. Everything about these modern machines is gigantic.

I have used the examples of mining operations in illustrating the gigantism of Homo Colossus deliberately. To feed its enormous appetites has required that we dig deep into finite stocks of minerals, extracting and using up resources that might otherwise have been left for prosperity. Of all the occult substances found deep in the nether regions of the earth, none can hold a candle to the devil's blood: oil. Here, in decaying carbon material, Homo Colossus found its preferred food. Ecologists would classify Homo Colossus as a detritus ecosystem since these are the ecosystems that feed on decaying carbon materials. These are the ecosystems that feed off dead biomass, breaking down the complex arrangements of molecules that sustain life and releasing their elements back into the cycles of material flow. These ecosystems rule the compost heap and the graveyard. These are also the ecosystems that are prone to population overshoot and collapse.

The Monster in Our Midst

It is time to take a step back from the scientific models and ask ourselves, what does all this mean? What do these ecological concepts mean for a compassionate, caring individual, our families and our societies?

I think most people fundamentally want to know they are doing good by the world. They want what is best for their children and loved ones. A consensus has been built up that business as usual was leading all of us to a good place. Progress was hard work but the sacrifices were worth it; from cutting down old growth forest to build a new settlement, to the second job to help the first child of this family get through college. The difficult unequal social arrangements of the modern world have been easily accepted largely because the promise was implied that if we could just lift the standard of living for the rich high enough, the process would inevitably improve lives for the poorest of peoples as well. The justification for consumerism as culture is that only through development can the desperate suffering of the third world be improved. If they keep working at it, the almost unspoken justification for our consumer lifestyles runs, they will someday be just like us.

In practice the third world is strapped with debt to first world banks for expensive first world infrastructure projects built by first world companies. Since the poorer country is able to borrow only so much, the rich governments of the world "give" them aid dollars with the stipulation that they can only be spent on "infrastructure improvement" projects. *Confessions of an Economic Hit Man* by John Perkins gives a small peak into what is going on. This is how the wealth-pump of empire works, pumping wealth from the peripheries into the imperial core. The companies and banks of the over-developed world profit but the question remains, did the recipient country benefit as well? There is no simple answer. In some cases the graft runs rampant and the whole adventure is one of abuses to land, animals and people. In other cases things are properly constructed but because the rest of the supporting infrastructure is missing the benefits are much less than what had been promised. In other cases real benefit is given, improving the daily lives of the poor by creating hospitals, education, sanitation, and widespread literacy.

In Western culture with its root in the ideal of Christian charity, crass consumerism was considered no more than the outer form of something more important and virtuous. Consumerism funded the cornucopia of technological progress. There has been every reason to believe in this secular god, progress. The so-called Green Revolution did manage to feed many additional billions of people since it began a few decades ago. We set foot on the moon, scanned the brain, and shared it all with TV, radio, computers, and the Internet – it is all very real and impressive. We are grateful and expect it will continue. Yet just here is the rub. Ecology states unequivocally, 'No, this pattern of consumption will not continue.' It would take multiple earths to bring the underdeveloped world to the state of industrialization found in the overdeveloped world. They say it would take three earths if all were to live as Europeans, five earths if the target lifestyle is that found in the United States. The implied promise behind the whole consumer shtick is shown to be bust, a sheer impossibility on a planet of more than seven billion people. Holding out the hope that someday, somehow it is still going to happen is now nothing more than cold cruelty.

There is a meaningfully sustainable degree of technology that we can all hope future generations may find. Today what we see are the deprivations of those suffering from not having enough infrastructure and technology to lead decent human lives at one end and those suffering total domination by the machine at the other. There must be a middle way of using appropriate technology sized to a human scale if our wisdom can find it, a way to avoid the extremes of underdevelopment and overdevelopment.

Looking around us today, this is not the future we ordered when we began this industrialized consumerism; collapsed fisheries and dead zones haunt our oceans, the land is scarred with cesspools of heavy metals and hot nuclear wastes, even the very air we breathe has become toxic to the stability of climate, all the while our consumerism is causing the sixth mass extinction - ghoulishly wiping out an estimated 200 species every day above the background extinction rate.

Looking around today many good people are questioning the formerly unquestioned foundation on which this whole thing depends: that human progress is only technological progress. This vision was sold to us by those who profit from our entrapment. I drive a car, I contribute to global warming. I buy food from a chain grocery store, I contribute to topsoil loss. On and on it goes, right on through the litany of horrors that is a typical day in the overdeveloped world when seen through the eyes of critical ecological analysis.

What happens to a culture that loses its most fundamental belief? When the justification for the blood, sweat and tears of generations no longer works? I certainly do not but do see the process playing out all around us. Seeing through the norms of the overdeveloped world's culture can be a most unpleasant waking up.

To retain strength and to honor that which is decent in human beings is the challenge. It is important to distinguish between the bitterness needed for dismantling Homo Colossus from any dispersion we might be tempted to cast on Homo Sapiens for having fallen into the trap made by our machines. Given the chance I am pretty sure the mosquitoes and the lions, the elephants and the Blue-footed Booby would have used the energy bonanza in a way not all that different than we did. Perhaps, as the Native Americans teach, we are among the youngest of our animal brothers and sisters: still intoxicated with the enthusiasms of youth and with plenty left to learn.

3.
Waking Up

Industrialized for all the right reasons

It was not that long ago that as a global society we were quite convinced that we were entering the space age. The excitement around placing a man on the moon restored hope in our technological progress. That hope had suffered serious setbacks with the invention of the death camps and nuclear weapons in WWII. Given the chance, people were quick to put the darkness of the past behind them. Millions of people flocked to read the visions of life among the stars the newly popular science fiction authors explored, and to see the transformation of everyday life right here on earth into the wonder land of technological convenience and leisure on display in the World Fairs of the 50s and 60s. In the United States suburbia was born, fueled by the GI Bill.

Walt Disney understood society's psychological need to celebrate childhood and innocence after the harrowing half century of wars and economic depression. He invented the first theme park as a place set apart from the dog-eat-dog world of economic, religious, political and military competition. It was a place devoted to delighting our children but without the edgy carnies and freak shows of the traveling circus and boardwalks. The survivors of the harrowing half century simply wanted to create somewhere safe for themselves and their families. People needed to believe they just might have a chance to die a regular old death, instead of one of the many horrific ones that had been inflicted on the world during the previous decades. After so much human evil had darkened man's image of himself, there was a real need to remake that image. Walt Disney understood the power of childhood's joy to do just that.

Waking Up

In Disneyland the parents were also given a golden ticket into dreams by being given a vision of a safe and comforting technology. By the 1960s the park featured the World of Tomorrow where the rising expectation that our technology could indeed be tamed was everywhere on display. Our technology would deliver to the human race a wonderful tomorrow among the stars. In the midst of the jet car rides and rocket ships of the World of Tomorrow there was the attraction where mom and dad could view their home of the future, the Carousel of Progress. Today we understand the corporate sponsorship and with our more cynical eyes understand how the notion of 1960s style consumerism was not just put on display but being sold to a public whose fears were easily manipulated. The Carousel of Progress had its part to play in these things but back then, when the home of the future was first introduced, it was a material incarnation of the idea that tomorrow could be better. As the carousel carried the theme park passengers around in a slow moving circle, they saw, as if in the center of a mandala, the evolution of technology from the 1900s, to 1920, 1940 and on to the wonders of the 21st century as they were anticipated in the late 1960s. The rooms were clean and the food was plentiful and, critically, the family was shown celebrating a good life made better by technology. Everyone from the household pet dog to grandfather were shown in a surprisingly ordinary way; relaxed and happy. The contrast with the bombed out rubble of Europe needed no explicit commentary.

Some had dared to dream reason and technology just might be able to bring us a world where we need not fear war, starvation and disease. Some had even hoped that through these means mankind would finally achieve abundance for all. After the hell that they had been through there was a widespread belief that this was the only way humanity might ever put an end to the scourge of war itself. It is all too easy for us, this far removed from those events, to fail to appreciate the devastating psychological toll they took on everyone. A wound was cut into the collective mind and no one was immune from its toxicity. Only by appreciating the zeitgeist might we understand what was happening as the consumer culture was being born.

It is hard for us to remember these dreams of an innocent consumerism, hard for us to give them any credence, but that does not mean they did not inspire our forefathers and fore-mothers. It was a good dream. It built on some solid features found in the power of human reasoning and some of the most noble of our virtues.

Mass advertising and public relations were invented to share the enticing cornucopia of consumer marvels the post-WWII factories were being re-tooled to deliver. The jet-set heading to the space age seemed just around the corner for an expanding middle class. I can imagine thousands of engineers were happy to turn their skills away from designing tanks and bombs and start using them to design products for the home and the needs of business instead. After so much bloodshed and horror it must have seemed a new world was dawning.

No one intentionally sat down at their drafting table to design a technological infrastructure that would destroy the vitality of the natural world.

Meanwhile advances in medicine were changing the very landscape of human suffering. Diseases and infections that had brought crippling disabilities to countless of our ancestor's and their family members were eradicated. Many sources of certain death were tamed and the population grew exponentially as a result. Again, no one intentionally sat down at the pharmaceutical lab desk to design a medical infrastructure that would destroy the vitality of the natural world. Yet, all those new mouths needed to be fed. Other highly trained, highly motivated individuals took the task to heart. Again, no one intentionally sat down to redesign agriculture and provide farmers with a Green Revolution that would destroy the vitality of the natural world. But they did.

We are angry. The situation we find ourselves in makes us want to blame others for our ecological predicament. It is easy to blame the greed of corporations and the villainy of their research departments, but that is not fair to the facts. When the choices were made that became the consumer society, the overwhelming majority of the population had no reason to think the activities of mankind would grow so enormous they would alter the very planetary systems on which all life depends. After all, those systems in the aggregate had remained inviolate for the many thousands of years of recorded history. Environments here and there were known to have been decimated; that there were once cedars in Lebanon was common knowledge. The whole earth itself, however, had never been threatened throughout the rise and fall of numerous nations and civilizations. I think it is important to remember that those who first laid the features of the trends that would grow up to eventually strangle us were not aware of this as anything but the remotest of possibilities.

All that changed in the 1970s with the publication of the MIT report sponsored by the Club of Rome, *Limits to Growth*. It was the ecological shot heard around the world. It was soon translated into 37 languages and sold an estimated 12 million copies - at a time when the total population of the earth was approximately 3.8 billion. It remains the top selling book dealing with ecological issues of all time. I believe these facts justify the claim that the world was hitherto put on notice that industrial civilization's ecological impacts were impossible to sustain. I think it is important to allow for this ecological naivety, which we have been discussing, among those dedicated people whose ingenuity and hard work have brought us so many of the wonderful things our modern world provides. Only then can we possibly comprehend the social and psychological ramifications of what happened next. When the *Limits to Growth* study put us on notice that our fossil fueled industrialized civilization was most likely to breakdown in the next century, the initial reaction was many-faceted and, my hypothesis runs, critical to understand if we are to comprehend what most ails us today.

What is it that most ails us? I suggest it is a lack of rational response to this ecological news. Our industrialized societies are unable to create a response that could be considered rationally proportionate to the threats that are described by ecological scientists as being immanent. If we were able to change this behavior the prospects for the future would be much brighter than they look to be today.

When the call came out to put a man on the moon "before this decade is out" Americans responded. Tens of thousands of individuals challenged themselves with the toughest science and math classes so they would be prepared to join the ranks of those working for the astronauts. Each one of these individuals were inspired enough by the vision being offered to set aside many opportunities for easier, more immediate gains for the sake of more difficult, but more fulfilling, long term options. It is said, "Where there is no vision the people perish."[7] This is what not perishing looked like. We need to find our way towards a realistic but no less powerful vision of what the dawning ecological age offers us. We did indeed place a man on the moon before the decade was out, and it was a most astonishing achievement for Homo Sapiens. We learned a lot about space. We learned a lot about ourselves. We call it the space age though perhaps, looking back many decades since that fateful day, it might better be called the earth age. What we discovered when we set our foot on the dry, dusty, dead moon was just how precious the small blue and white globe we call home really is. What most took us aback were not the images of the moon - it was the image of earth hanging like a jewel against the black of space.

It was the birth of knowing, with photographic evidence, that life on planet earth is wholly interdependent. It was the picture of the fragility of our own planet, lightly rimmed in blue against the vast cold reaches of interstellar space, that was destined to have the most profound impact on the collective experience of our species. Suddenly, in the blink of an eye in evolutionary time, our species had grasped both the power to divide the very glue that holds the universe together in atomic fission, and the power to view where the whole of known life has taken place in a glance. That view of the whole could only be seen from the heavens. Our species saw it with our own eyes. This was the context in which *Limits to Growth*, leveraging our new thinking machines, the massive early computers at

Waking Up

MIT, delivered the bad news. This was the cultural context for the study that shook the industrialized world. Unquestioned cultural assumptions were torn asunder when what had seemed the right thing to do had grown into something unsustainable. We knew technology had a dark side, we had been exposed to it in the camps and weapons of the war years and the grueling lack of dignity the suffering of the Great Depression had caused. Now it seemed that with this study coming out of MIT as though our machines had spoken like oracles. What they had to say seemed to be their final betrayal.

There is a Faustian tale here, full of heroics and the fires of Blake's satanic mills. Of course, this seeming betrayal at the hands of our technology was really the projection of our own unbridled greed and violence. Still, once it looked as though our machines had betrayed us the door was opened for our leaders to betray us as well, and themselves. Since the 1970s when the *Limits to Growth* study was first announced a whole library of documentation has been gathered as evidence against consumerism. We have long ago lost our innocence. Culturally, we took on the schizophrenic stance of both knowing, and pretending not to know, that our lifestyles were unsustainable and causing serious damage.

Our societies have been suffering from a serious case of cognitive dissonance as a direct consequence. An element of the conman began to grow in our public life like never before. The sham and hustle-hustle grew larger until the integrity of Wall Street and government became ever more corrupted under its influence. The new times called for a new type of leader. To maintain business as usual, to in fact thrive on it, required someone who knew seeking endless growth in subservience to the cult of shopping was killing the planetary ecosystems on which all life relies, yet would be willing to do it anyway. These new leaders cannot be compared with those of previous generations simply because they were acting at a time in which all educated individuals on the earth knew the real score. This is true not just of our leaders. The people allowing the con to proceed have not escaped cognitive dissonance either. Populations can look the other way but passing the buck doesn't change a single ecological fact.

Our fear of suffering the horrible technological deaths we know countries can now deliver with the flip of a switch is what has made it so very difficult to begin an honest conversation about where we are as a species, and where we need to go if we are to minimize the suffering ecologists assure us is heading our way. It would serve us well to learn to suffer honorably again. We do so when we learn to suffer for the sake of something greater than ourselves. We need to speak clearly: there is no winning the war against nature, there is a future in learning to use our ingenuity to serve nature.

Since the 70s our position has changed. Mindful Ecology insists that it is here that we arrive at the crux of the matter. Business as usual is ethically, morally, and scientifically bankrupt. Our corporate leaders are unable to address these things because they are hired to maintain the profits of those benefiting from the existing state of affairs. Costs to future generations do not enter into their calculations of quarterly profits and, it is important to recall, no consideration other than quarterly profits are legally allowed to influence their decisions anyway. Our governmental leaders remain a wild card but so far do not provide much evidence that we can expect any rational and honest appraisals of our options from that quarter either.

This is where individuals are called to stand up and make their voices heard, to stand up and let their lives witness to values other than the death trap ones being, literally, sold to us. There is a very real liberation to be claimed by those with the courage and wisdom to find it. The majority of what ails existing societies in our misuse of technology is very much a man-made phenomenon. It is not inherent in technology itself. It is in the choices of what kinds of technologies are developed and how the distributions of goods are designed that determine whether or not any given application of our scientific knowledge is in harmony with our best interests. Socially, we could turn away from the waste generating economy, delusional with unsustainable ideas of endless growth, and turn towards a

Waking Up

steady state economy oriented towards the production of fewer higher quality items. Socially, we could turn away from the constant discontent sewn in our advertising driven communications so that we remain always hungry for more, and instead encourage the use of less energy, stuff, and stimulation in the pursuit of high quality lives filled with contentment. We could restore the vision of a safe world, one in which we seek the good by restoring the wise vision of our elders' teachings that things do not buy happiness. With things we produce returned to their proper place, socially or individually, it becomes possible to pursue a cultural purpose befitting the inborn nobility and dignity of Homo Sapiens. We can choose to try and learn how to shape social environments which encourage living harmoniously with one another and the earth.

Society could have done these things half a century ago when we first learned of the dismal fate awaiting the industrial world. We did not make those changes and still have not really started them at any appreciable scale. This is the crux of the matter. Now it is late, only individuals have any power in the equation that will determine whether viable alternatives exist on the other side of the breakdown of business as usual. We need to wake up to that fact - our existing leaders cannot and will not provide a response proportionate to the crisis. What then is to be done?

When the ecological vision is found, the person again finds the strength of will to set aside short term gains for the sake of long term ones. This lost art comes naturally to one who has learned to seep themselves in contemplations of deep time and deep space. This is the one thing our existing social arrangements make it impossible for us to do collectively just now. This is a problem. We need to learn to act today in the interest of our long term tomorrow socially too. We need to learn how to do it if we are going to have any chance of minimizing the unnecessary suffering heading our way. However, we as individuals are not slave to what the collective chooses to do or not to do. We are individuals alive at a time of unprecedented danger both mentally and physically. This is a time marked by ubiquitous magical thinking in Disneyland-like fantasies which are acting as a veneer over our unprecedented exposure to pornographic sadism. All the while our leaders are encouraging seriously delusional ideas about our capability as a species to survive without intact ecosystems. The way these collective delusions will be unmasked is by individuals seeing through them. In this case it is relatively easy to see that the emperor has no clothes. The cognitive dissonance - knowing the oceans are dying even as you drive to the store to buy fish for dinner tonight - has become all but unbearable. Something is going to give and we as individuals should do all we can to prepare for that. Once you are convinced about the way the probabilities are pointing you will have a choice to lay up your own bombs or your own wisdom. How will that choice work out for you? Will you suffer the loss of false certainties for the sake of the truth, or will you suffer from the false certainties themselves?

Mindful ecology is not meant to invoke the ideas of other worldly enlightenments but to turn our attention to the environment right in front of our noses. We take courage and reason as our guides, along with a heart broken with compassion for the suffering our ignorant ways are causing day in and day out for so many sentient beings. Once you see that clearly, you also see it need not be this way. We are seeking to develop the skill of the steady gaze in this time of fly by night salvations. These are dangerous times but this just means our fight to wake ourselves up to the reality of reality is all that more inspired. Many of our neighbors, friends, loved ones and companions currently believe a whole lot of B.S. First light your candle and then seek to light theirs. It is better than cursing the darkness.

1970s Wake Up Call

Once a person really grasps the dire nature of the ecological reality around us it is hard not to wonder why the generation of adults in the 1970s did not do more to change the trends they had learned

Waking Up

were leading modern civilization to disaster. It is easy to blame but that blinds us to what we need to understand to recover our power now, in this present time. We need to look with compassion that understands what it was that was happening to the human psyche.

This decade saw the concern about the health of earth's ecology become a shaper of human events for the first time. Mankind had seen the earth from space. It is hard to underestimate the impact this made on us as a species. In this decade Love Canal caught fire and coal power plant produced acid rain began decimating the forests of North America. The citizens of the United States awoke to the fact that they had polluted their own nest. The Environmental Protection Agency was created in 1970. A flurry of activities was undertaken, including the first Earth Day and the publication of the *Limits to Growth* study. The public insisted that something needed to be done. President Nixon, a republican, first extended the 1963 Clean Air Act in 1970 and then, under continuing public pressure, expanded the 1948 Federal Water Pollution Control Act in 1972, creating what became known as the Clean Water Act. The large environmental organizations i.e. the Audubon Society, the World Wildlife Fund, and Greenpeace grew from seeds first planted in that decade. They would continue to lobby Washington and keep ecological issues in the public eye from that day forward.

The record of results since that decade is spotty at best. With hindsight we can see that cleaning up America was achieved to some degree, but at the cost of moving the heavy industries that caused the bulk of those pollutions offshore. Those who argue that today the United States needs to bring manufacturing and heavy industry back are right to couple that policy with gutting the environmental laws, these things are of a piece.

When humanity saw the picture of the earth from space it looked fragile; there was only one and we were alone, on our own out here. Every generation has had some people who have feared an apocalyptic end of the world was just around the corner. Now this picture of our earth's absolute independence in space appears as if to underscore that the human race really is on its own out here. Our race had recently come through concentration camps, nuclear weapons and economic collapse. It would not have been hard to fear even worse things than these were heading our way, in fact, it looked probable. When the news came out of MIT that our civilization was likely to overshoot our resources and overwhelm the natural world on which we depend, well, just imagine it. People cried. I was 10 years old and remember my neighbor's father, an old Navy man, talking on and on about it. Before long he grew quiet but he carried a look of defeat and sadness that had not been there before. As his attention turned to Watergate, the other bad news of 1972, his hair quickly turned gray and not long after that his heart gave out.

It was in 1968 the Club of Rome asked for a study of the world problematique using system science and computation. The resulting study was completed about the time the David Bowie song *Five Years* was written in 1971.[8] These things were in the air. The opening verse of that song states clearly "earth is really dying." In interviews with the *Rolling Stones* and William Burroughs, Bowie explained this was because "It has been announced that the world will end because of a lack of natural resources."[9]

What Bowie the artist was able to see, I propose, is how this type of news was changing people. He identified the archetypes that would accompany us on our journey bearing this "terrible news." In the interview he mentions how it is a time in which adults are out of touch with reality and the children are left to plunder anything they think they want. I look around at the dissolution of the barriers between adult and child material in our society and think his couple of sentences from the early 1970s capture the state of things rather well.

For what it is worth I believe this vision into the resource restrained future stayed with Bowie the rest of his days. *Blackstar* takes up the theme and in his penultimate work, appearing after a 10 year hiatus from public music making, it animates *Where Are We Now?* ("just walking the dead" the song answers). The video he released with that song in 2013 artistically captures faces frozen with anxious eyes watching as time passes by.[10] The historic film included in the video is showing Berlin in these fateful 1970s.

"Pushing through the market square,
so many mothers sighing. . ."
David Bowie, *Five Years*

(Respecting copyright I will assume my readers either know the rest of the lyrics to this song or can find them to read in conjunction with my comments below. To my ear it is almost the theme song of Mindful Ecology. I say almost because Louis Armstrong's *What a Wonderful World*[11] takes pride of place for me.)

The market square is of course where we wove the spell under which we grew from the sustainable human scale to the unsustainable Homo Colossus. It is pushy in here as the pie shrinks. I get an image of young mothers pushing strollers with their children in them, bleakly wondering how can they be happy for their children again? With a heartbreaking sigh they look on the reality, such a small amount of time left to cry in. What his hauntingly beautiful lyrics proceed to express for us is the emotional impact the news of limits to our growth has had on people. It hurts. It is true. There is no escape, what are we to do?

When this vision gets 'stuck in your eyes' everything you see in the normal world of struggling humans is touched. Everything we have worked so hard to achieve suddenly looks to be so very impermanent. We scan the built environment and the culture so intimately bound up with our identities, like the identity of Bowie once a boy fascinated with a toy from father, mother's electric iron and the wonder of TV. The consumer cornucopia is our embedded mind, our brain made over like a warehouse. These common, everyday things carry powerful emotional connotations in the unconscious mind. As he thinks about the loss of all these things he had taken for granted since his childhood, he suddenly realizes what this "terrible news" will also mean for all the people in the world. That it will touch billions and billions of us.

His attention turns to people: a world full of laughing, crying, struggling human beings with warm bodies in an endless variety. Some, like himself, destined to give form to the dreams of the many and others destined to be the many. The contemplative who has worked with compassion understands the truth, "I never thought I'd need so many people." We are the same. Equal.

A whole lot of people were going off their heads as the 60s dreams of the Age of Aquarius were shattered in the dark 70s. I read in the lyrics an accurate description of what archetypal psychology knows as the dark mother. Those mothers pushing their baby strollers through the market square sighing sometimes snap. How could it be that this most wonderful and beautiful baby of mine is the source of so much darkness and pain in our over-populated world? Going off her head, the unconscious rage at the unfairness of it all was given free reign. What happens next is just as Jungian thought would expect; sometimes the only way to beat a monster is to invoke a bigger monster. A compensating darkness rose to protect the survival of the battered child. The shadow can be a person's protector. The theme is common enough. A recent example is the movie A Monster Calls which deftly handles this material. Missing dads, dark moms, and monsters make up the heady concoction in consumer capitalism's bitter cup.

Next evocatively, have we not all broken our arms saluting the military-industrial complex one too many times, while staring with mono-vision at the automobile as the summit of industrial wealth? When Law and Order are put in service of the True Believers it makes society's outcasts sick with fear and disgust. The sad news invokes not only the dark mother. The dark father puts in an appearance as well as patriarchy's physical violence lends its services to spiritual violence, causing physical sickness among the broken and abused they leave in their wake. It is an odd justice that locks up the addicts and prostitutes and lets the pedophiles, rapists, and terrorists who created them go free. "Take a look at the lawman, beating up the wrong guy" David Bowie sings in his 1971 song *Is There Life on Mars?*

Waking Up

Our love for our children is stronger than our fear - even now. Our children bear our face, our race, the way that we talk. They speak directly to our hearts. Overshoot is not left as an unfelt abstraction. The child is recognized as precious, even by all the darkened mothers and fathers. Evolutionary blood, sweat, and tears have made my child just who and what he or she is. The deepest parental wish is that one's child will be able to balance their physical, emotional and mental lives so that they can live a good life, and that the world they encounter will provide the stability and support a good life requires. All this, and more, is in that quintessential parental gesture: offering a steadying hand for our child's grasping finger as they take their first trembling steps. "I want you to walk." Let us pray this spirit of help and nurturing will become more evident in our culture.

The people of the industrial world were surprised by the terrible news. It pains the brain but once you know, you cannot forget. It gets stuck in your eyes; everywhere you look, you see it.

What, exactly, was an individual to do? This is the important question we need to ask. Some of these young mothers were terrified, some of the young fathers overcome with anomie. The will to live lashed out with a lion's roar; the 70s were also known as the decade of excess in sex, drugs and rock and roll. It was time to change the world. The problem was that the innocence of the 60s was lost midst lives broken by hard drugs, thoughtless promiscuity, and the traumatic violence of the Vietnam war. A war, we need to recall, that was being televised in graphic detail which proved so traumatic it is something, it is worth pointing out, our leaders no longer permit.

Mindful Ecology suggests that ever since that time, this special decade of the 70s, the leaders and citizens of the industrialized nations have understood our way of life was unsustainable. Many people have wailed against what they thought was ignorance on the part of the public concerning these ecological matters of life and death. I do not think this was the case. We are not unaware of our circumstance.

We know.

We know what is happening. What, exactly, is an individual to do? There is no action an individual can take that even begins to seem proportionate to the size of the problem. That was the truth that the generation of the 1970s were the first human beings to grapple with. A kind of 'we know but we are not going to talk about it' norm took over. Sociologically the conversations around the end of growth between friends and neighbors became taboo. The media followed suit and went on to its next set of titillations, pretending as if business as usual had not just had the rug pulled out from under it. We have been living in the cognitive dissonance ever since. It is an extremely painful dissonance because it involves some of the deepest longings of our emotional nature. Our hearts are wrapped up around our hope for our children. Call it our evolutionary imperative if you must. Love is the engine of everything we social primates do as the most biologically complex of all the mammals on earth. In the dark despair of our nightmares we ask ourselves what if my descendants are not able to fall in love, to chase the spring into summer and enjoy the blue sky. To even consider oneself guilty of committing such a crime is a horrendous burden. To do so for something as trivial as driving a car is perverted. We chose to repress our awareness of the jagged edges of our daily lives as best we could. What else could we do? We had to get on with the business of living. The consumer culture was unprepared for seeking wisdom in its suffering.

We are still caught on the horns of the same dilemma: what, exactly, is an individual to do? What exactly is a culture to do that discovers it has been wrong about so many things? Mindful Ecology is not an answer to these questions; it is the suggestion to stay with them. Repression of the painful truth did not serve us well. It is an open question whether or not it will prove to be necessary to repress public discussion of the true extent of our ecological mistakes all the way to some very bitter end. History will tell. What is unsustainable will not be sustained, of that we can be sure.

Waking Up

Regardless of what the society decides, individuals remain free to choose for themselves how they will approach the ecological reality of our situation. It is hard for a modern mind to approach a predicament, a problem for which there is no good solution, a problem that cannot be fixed. Everything about the high technology culture which giant industrialization has made possible embodies an ambition to conquer nature. To question that, to ask if that is one of the fundamentals that needs to be critically examined, to admit that whole approach is becoming a disastrous failure, such questions leave us in a psychologically uncomfortable space. Our minds are grounded in ideas they deem solid, this is an earthquake. There is plenty we can do to minimize the damage but there is nothing we can do to rescue the system itself. We need to find the courage to at least entertain this as a possibility. Otherwise it will continue to rule by haunting our individual lives from the shadows. The good news is that here and there people are reporting that by staying with the suffering involved in overcoming this particular cognitive dissonance, it transforms their lives. The changed values reveal a changed world because they have allowed biophillia to change their subjective experience of how they approach the world and all the sentient beings within it. They report it is the way to come alive again to the world as sacred.

What this means is that it is more or less a waste of time to try and convince people that there really is a problem. People know that, people everywhere feel the truth of it in their bones. The whole issue needs to be approached from another angle. This is the catch-22 that is keeping us entranced: there is no solution to the problems generated by our lifestyles to be found by changes within those lifestyles. It is changing the lifestyles themselves that is the solution. Alternative lifestyles to the consumerism of giant industrialization, however, are not allowed. And that is the rub. For as long as the necessities of life can only be acquired from the teats of the megamachine, it will remain imperative that the beast be kept alive. None of this is fundamentally true in the molecular world, the existence of oil fueled giant industrialization can be traced back just a few centuries and is unlikely to reach many more. The bridge from here to there is what concerned people all over the globe should be busying themselves about. Today, every day, there is some little thing we can do for the immediate environment we live in and for the mental environment we live in. Will it save the world? No. We need to get this through our heads. The world does not need saving. What does need to be preserved is the fabric of human culture and history, our liberty and our dignity. As events along the path of collapse continue to drive us all a bit crazier, these things are threatened on every side. So much is worth saving, so much needs to go and so much work is needed.

This is not a game. This is not a way to sell books. This is a frightening reality which so far the human species has not been able to deal with intelligently at all. It does not look like there is going to be a happy chapter at the end of this tale.

> "News guy cried when he told us,
> Earth was really dying.
> Wept so much his face was wet,
> Then I knew he was not lying."
> David Bowie, *Five Years*

The end of industrial civilization due to the over-extraction of resources and the over-production of wastes is creating a decimated biosphere. In the opening decades of the twenty-first century there is little left to discuss around the basic concerns first clearly expressed in the *Limits to Growth* study. The evidence that our societies chose to follow, and reap the consequences of, its dismal business-as-usual scenario is now everywhere; from the ocean and air absorption of CO_2 going off the charts, to the rate of new discovery of oil plunging into the abyss. We are heading into unprecedented territory at a frightening clip. We could be in for a very rough landing.

4.
Homo Colossus

"Man has been too arrogant in exaggerating the difference between himself and other creatures, between humanity and natural
history."
Overshoot, William Catton

Life Under Homo Colossus

Under the reign of Homo Colossus there is a tendency to
obscure key ecological relationships and their results. In
an Orwellian fashion we refer to the extraction of non-renewable
resources such as coal and oil as 'production.' Countries like the U.K.
claim to be meeting carbon reduction targets, when in fact much of
what they have accomplished relied on moving the polluting manufacturing industries on which they still rely to third world locations.
Pundits rail against the raw population numbers in underdeveloped
countries while ignoring the role of affluence and consumption patterns when discussing humanity's ecological footprint.

Living in the time of giantism in technology, transportation, communication, information, and resource throughput is living under
the reign of Homo Colossus. Homo Colossus consists of numerous
complex networks. When they are all working well together they deliver bread to the store shelves and the trains run on time. When the
slightest hiccup disrupts one of the networks, that oscillation can be
multiplied until larger, seemingly unrelated networks also fail. We
see this in a traffic jam started by a few distracted drivers gawking
at a fender bender on the side of the highway. Before the waveform
has completed its travels through that city's membranes there could
be a five mile backup on the highway, a five percent increase in traf-

Homo Colossus

fic accidents and an extra metric ton of carbon monoxide added to the atmosphere than there would have been during this morning commute had there not been a fender bender. Though this is a silly example we need to understand that often when the complex network of systems a city relies on for survival breaks down people die. Modernity is poised on the edge of the complexity knife with giant systems everywhere becoming more brittle and prone to disruption as the pressures of population and scarcity grow with each passing year.

The entropy attack on the American infrastructure is there for anyone to see who travels through the back roads of this country. The derelict factories, farms, abandoned suburbs and small towns are everywhere. When the traveler turns towards the mega city population centers, particularly if they stray into the inner city underbelly, there are all the signs of third world poverty and neglect. The systems that keep the food on the grocery store shelves, the potable water in the tap, the gasoline coming out of the pump and the rest of the essentials we all too easily take for granted are each baroque tapestries of interdependencies. Keeping the lights on involves a whole slew of engineering design, procedures, maintenance, oversight and cooperation all coordinated to a very high degree. These giant systems seem to be invulnerable until a hurricane Katrina comes slamming into them or the earth shakes and for weeks these critical systems suffer disruption. Individuals and families not blindsided by this country's happy talk might want to minimize their dependency on these systems.

The Age of Exuberance brought high hopes that are now being dashed. If disappointment tests the mettle of a man, as a society the overdeveloped world is finding out just what it is really made of. Life in the time of human overshoot is characterized by particular social stresses. There is an increased competition for a slice of the pie as the population of the middle class shrinks. Stalemate becomes more common in democracies as nations can no longer afford to pursue competing political goals. Along with the competition over the shrinking pie come violent antagonisms and along with overcrowding comes a fear of being redundant. All these sociological forces have their roots in ecological changes. Understanding this promotes a compassionate response and inoculates against catching the fevers of the demagogues.

Lacking ecological insight it is all too easy to blame a sinister cabal, the one percent, a hostile government, incompetent bureaucracy or corrupt politicians for our troubles. Not to say all these and more did not have a hand in creating the mess we are now in, they certainly did. Yet, they too were being played by the sea changes overshoot is bringing about. As the coming century brings adjustment and correction to the human race it will be all too easy to fall prey to the rhetoric of a Caesar selling us a story we want to hear; the American dream was stolen by evil forces, the promise of the enlightenment's improvement of the human being was perverted by scheming cabals in corporate boardrooms, the nations are again soaking in blood because of greedy warmongers... So what are the key ecological insights by which we might retain some sanity through these troubled times?

First and foremost ecology teaches that everything is interrelated, that there is a universal interdependence among all that lives. We share the same chemical basis, the same DNA information replication mechanisms and many of the same bio-physiological pathways that many of our most primitive ancestors used. Mankind can look to the causes and conditions of other life forms for lessons, recognizing there have been numerous precedents of biological communities undergoing succession, exuberance, overshoot and collapse. It is not that we are reaping the rewards of a fatal character flaw but are simply coming to the natural end of a natural process.

Perhaps the second critical insight ecology teaches is that everything is always changing. It always pays to ask 'and then what?' Biological communities change through the process of succession. As the population expands the community alters the very environment in which it evolved and paves the way for the next biological communities to succeed. By seeking to maximize GDP as our social goal we accelerate the process of succession whereby we undermine our

Homo Colossus

own habitat's capacity to support our species, or more accurately, Homo Colossus. And then what?

In my short list of key ecological concepts the third would be that ecosystems are open systems. All living things sustain themselves far from thermodynamic equilibrium by taking some substances from the environment and sending other substance back into it. These exchanges in turn influence other organisms, the outputs becoming inputs in a multitude of recycling cycles. If the chemical byproducts of one organism are toxic to another their relationship is antagonistic. Smog is antagonistic to trees. Note that ecological antagonism is impersonal; the drivers of the cars creating the smog do not hate the trees. Perhaps this is easier to see in a case where human concerns are not involved. Penicillin is antagonistic to a variety of disease causing bacteria; it is a 'pollutant' for the bacteria. Basically anytime the output of a life process seeking its sustenance becomes toxic to another life process the relationship is antagonistic. With this definition in mind it is not difficult to see how it applies to human situations. Animosity arises from interference of populations with one another, even when unintended. There can be valid reasons for antagonistic relationships between human groups without the need for villainy or human perversity. By not recognizing this, overcrowding can become cause for war. The frictions are exacerbated by not recognizing the actual causes and conditions.

As competition increases within an ecosystem there is a tendency towards what is called niche diversification; a sharper differentiation between communities occurs in how they use their habitats. When applying ecological principals to human society it helps to see the wide diversification of human endeavors as if it had created a set of sub-species. The jet setting one-percenter uses our planet very differently than an organic farmer. The butcher, baker and candlestick maker all have unique interactions with their respective habitats. Our modern societies host the result of centuries of social diversification that was built up during the Age of Exuberance. When there was plenty of cheap energy to go around the competition between Wall Street and Main Street remained friendly, the competition between the working class and the needy was colored by charitable ideals, and the competition between generations was a source of good natured pleading and prodding. Remove the abundance of cheap energy to sustain all these relationships and the competition grows more fierce. The ecological antagonism promotes the emotional antagonism.

In addition to increased competition, in *Overshoot* William Catton speculates that overcrowding also exacerbates a tendency to defend ourselves or our tribe from the fear of being redundant. Most lives in the overdeveloped world are dedicated to the purely anthropomorphic goal of maximizing money and protecting one's own. There is a gnawing suspicion that one's life is not contributing to any healthy trends nor is it made meaningful by being dedicated to working for what is greater than oneself or one's family. As the sense of overcrowding spreads, it dawns on each of us that any of us can be replaced by one of our competitors. He writes we are "in danger of being considered superfluous ... the plight of the unwanted child became potentially everyone's plight."[12] In reaction we raise our voices in an anxious attempt to assure ourselves that redundancy applies only to other people. Much radical social activism is implicitly saying 'it is you, not us, that is superfluous!' with each group adding their favorite vindictive snarl words.

The holocaust could prove to be an ominous prelude to what happens when one group declares another to be superfluous, redundant.

Catton's mention of the unwanted child is darkly prophetic. What do the tragic shootings in schools all across America indicate about the social health of the country? Multiple causes are no doubt in play and any generalization is bound to be a simplification that cannot apply in every case but these caveats should not make us afraid to draw some conclusions. Why would children feel the need to eliminate one another? Might they sense that the future bearing down on them is one of fewer resources where increased crowding and competition shape their destiny? Murder is one way to assert that you are the one life can do without, not me. In a culture willing to burden future generations with un-payable levels of debt, depleted natural resources, and polluted water, air and land is it any surprise that the normal adolescent pain of rejection turns at times into the full tragedy of a violent outburst?

Life under Homo Colossus has perverted society, reshaping it all out of proportion to the human scale. The gigantism of our globalization assures its influence is felt planet wide. On a planet that is not growing larger increasing our numbers is bound to increase the level of antagonism, completion and conflict, all of which will be exacerbated by the desire of the overdeveloped to hang on to what they have and the underdeveloped to achieve full industrialization. A paper was recently published in the Proceedings of the National Academy of Sciences with the dismal news that, as the BBC headline had it, *Population controls will not solve environmental issues.*[13] A team of researchers investigated numerous fertility restriction scenarios to examine what impact they would have on total global population. This is an important question since voluntarily curbing fertility rates is part of the more humane ways of bringing our societies back to some form of sustainability. What they found in the study is sobering. The UN released population projections based on data up to 2012 and a Bayesian analysis. It found there is an 80% probability that world population will increase to between 9.6 and 12.3 billion in 2100. This new research found that restricting fertility is not sufficient to change these population projections substantially. Specifically if the whole world adopted China's one child policy in 2100 the total would still be between five and ten billion. If there were a global catastrophe mid-century, a war or pandemic that killed two billion people, it would barely make a dent in the trajectory over the next 100 years. We could still expect eight and half billion by 2100.

The take away from this research is very pertinent. While reducing population is the only long term relief from the pressure Homo Colossus is putting on the planet, there is such a large demographic momentum built up that nothing is likely to change the course of the next century. Curbing our numbers will not help us deal with the environmental problems we are confronting in the short term.

Most of the mainstream ecological pundits recommend a rapid transition to lower fertility rates as the best way of dealing with the increased impact of our rising affluence and consumption. This is said to be the way towards sustainability. It is a welcome solution in which the consumer paradise can continue, the economics of continual growth do not need to be fundamentally reworked and green corporations pursuing business as usual in the free markets are said to be sufficient for dealing with the ecological crises. This type of research puts these dated ideas out to pasture. The only lever that remains for actually lessening the human impact on the environment on a timeline that will make a difference is to cut back on our consumption. This is not an easy sell since for most people a diminishing affluence brings with it a diminished sense of identity as well.

The End of Homo Colossus

Homo Colossus is not long for this earth. With an appetite not even 10 earths could satisfy, soon this beast will starve to death. It will not be pretty, like a junkie cut off from their supply. We are talking here about the hard reality of ecological limits: consumption reduces the remaining stock of non-renewable resources. This is only common sense. The consequence is that there are limits to the number of non-renewable resources mankind will be capable of accessing as time proceeds.

Few subjects have suffered obfuscation by spin doctors more than the idea that there are limits to growth. The idea is so threatening to economics with its debt based fiat money that loud and pervasive voices work overtime to assure investors worldwide that there is really nothing amiss in the pursuit of unending growth on a finite planet. Our subject is the role that limitations play in ecological science and how that spells the end of Homo Colossus. Due to the

Homo Colossus

confusion deliberately propagated around limits, the first task is to take some cognitive garbage out to the compost heap.

An argument could be made that modern science was born and continues to be a powerful means of inquiry through a proper appreciation of limits. The calculus provides the mathematical tools for many of the most fundamental theories across a wide swath of the sciences from physics to evolution. With the mathematical tools of the calculus we are able to capture the rates at which things change. In a universe in which all things are constantly changing, the value of such a tool is obvious. In the calculus the mind numbing subject of infinity is tamed. The limit is one of the core concepts of the calculus. A favorite illustration of this limit concept is as an answer to Zeno's paradox. In a race between a tortoise and a hare where the tortoise is given a head start, the paradox runs, the hare can never reach the finish line. Why? Because before the hare can reach the line he must go half way, but then he must go half way through the remaining distance again. And half way through that remaining distance, ad infinitum. This is a logical conundrum, a time bomb hiding in our maths. With the calculus we are able to prove the hare can indeed cross the finish line, to the great relief of racing fans everywhere, by saying the hare approaches the finish line *in the limit*.

The maths we all learned started with arithmetic where sets of static things provide most of the mental models needed for its comprehension. The operations of addition, subtraction, division and multiplication are often modeled with a set of colored blocks in classrooms around the world. Who doesn't recall that proverbial set of apples our teachers went on and on about as we visualized giving and receiving some from our friends or slicing them into fractions? Most young students understand apple word problems fairly well. It is with the introduction of algebra that the first wave of math phobia strikes. Sure sometimes we want oranges instead of apples, but why call it x?

For many young people, who cannot help but notice that most of the really important issues of their lives concern ever changing qualities, the picture of relationships among static quantities algebra provides seems alien, of no consequence. I imagine the perennial complaint, 'but how will I ever use any of this in my real life?' was probably first spoken somewhere in ancient Persia right about the time algebra was invented. In my experience it is a shame really that so few make it across the algebra bridge into the calculus, since it is in the calculus that all those fiddly and seemingly arbitrary rules found in maths begin to all fit together. It is also when one of the most important intellectual streams of our cultural inheritance is transmitted to an individual: our sciences.

Science observes that all things are seen to be on trajectories that are inevitably thwarted. Enumerating a few examples reads like a who's who of scientific discovery. Evolution – an animal species multiplies but it does not fill the whole earth with its offspring, something limits its reproduction potential. Dynamics – a body in straight line motion tends to stay in motion but on earth friction always slows it and other forces divert it from its path. Cosmology – there is an absolute speed limit in the universe, the speed of light as per the theory of relativity. Geology – there is a limit to the pressure that can build up between continental plates until earthquakes occur, there is a limit to the force the crust of the earth can suppress before a volcano erupts. I could go on but the point has been made; scientific knowledge is very often carved out of our ignorance by the recognition of the factors that limit processes. Each of these examples and many more are embodied in theories that have mathematical models at their heart, models built using the calculus.

This is the larger background required to honestly asses the role of limits in ecology. I don't believe someone must understand the calculus to benefit from a mindful practice centered around ecology but it is important to recognize the role that mathematical models play in science generally (and why), so that when examining the theories and evidence in ecology a sufficient understanding is brought to bear. At the risk of oversimplifying, it could be said that when the sciences fit equations to data they have a small family of curves they can draw on to do the work. Learning to think in curves is a worthy addition to the cognitive tool-belt of anyone who wants to understand the modern world. Learning to think in curves is different than being able to manipulate the equations, the recipes, that create the curves. We are not all called to become mathematicians yet each of us can lay claim to understand our scientific inheritance if we are willing to do the work. This educated understanding sees a critical commonality: that everywhere throughout the natural world exponential curves meeting limits are producing the sigmoid and normal curves that characterize our universe. This is a key scientific insight into behaviors as wide ranging as atomic structure, differential reproduction and the curvature of space-time.

In ecological field work researchers try to identify the differences that make a difference. The subject of limits is central to the ecological sciences for this is often how the environment induces its selection pressure. The 'operationally significant' factors that control the abundance and dispersion of a community stand out from the buzzing jungle of details where it is difficult to tell if one thing is more important than another or not. To get a handle on the survival characteristics of the species-environment interaction we ask what limits its growth? For a field of corn it could be the availability of phosphorous, for a Petri dish of yeast the nutrient sugars could be the limiting factor. Of all the many, many elements a biological community needs to survive and reproduce there are typically one or two that are in short supply. The limiting element acts as a brake on the growth potential of the biological community.

In 1840 Justus Liebig expounded this principal that the availability of a limiting resource controls an environment's carrying capacity. It is known as Liebig's law of the minimum. Fertilizer is our solution to this limit problem in our crop growing efforts. Fertilizer is designed to provide just those elements that are in short supply so the harvest can produce its maximum yield.

In addition to the limit brought by the minimum critical factor ecology recognizes a second family of limitations. This "law" of the limits of tolerance was included in the work of V.E. Shelford in 1913. This is the set of limits around what living things and their environments find tolerable. Not only too little can be a limiting factor but also too much. Life is very sensitive to numerous boundaries which it cannot violate and remain viable. Temperature, salinity, and toxicity are a few of the better known. Mammals, an example particularly relevant to ourselves, exist within a narrow band of temperatures, maintain internal PH levels and must consistently remove waste products running the gamut from dead cells to fecal matter.

There are a few details worth pointing out. Organisms might have a wide range of tolerance for some factors but very narrow for others; I can drink a wide range of water volumes in a day and survive but can't eat too many strychnine cookies. When conditions are not optimal in one factor the limits of tolerance in other factors may become reduced; a meadow low in nitrogen needs more water to fend off drought. The period of reproduction is generally the most sensitive to limits; the seeds, eggs, embryos, and larvae cannot withstand the more extreme conditions an adult of the species could. Each of these details is worth some contemplation to tease out how they play out in both natural and human history as well as how they might contribute to the overall shape of the future.

No one knows when the limit to the giantism of Homo Colossus is going to be found. Will it be next year, five years from now, fifty, one hundred? No one knows which crucial element will not meet its supply without a viable substitute or which might move the habitat beyond our limits of tolerance. The way to investigate the issue, as we learned looking at the calculus, is to examine the rates at which resources are being used, pollutions are being produced and populations are growing. An uncomfortably large family of candidates confronts the researcher. Most of my readers will already be familiar with many of them, still, just to assure we are on the same page here: it is estimated the U.S. loss of topsoil is 10 times faster than it can be replaced; global arable land loss is 30 to 35 times the historical rate; species loss is estimated at 1,000 to 10,000 times the background rate; ocean acidification is increasing at the fastest rate in 300 million years; etc. Note that some of these are minimum limits

Homo Colossus

(lack of required topsoil) and others deal with tolerances (how much pollution the oceans can take).

In the late 1960s a team at MIT decided to use these tools to examine the overall shape of the future of industrial civilization. They dissected Homo Colossus. They created a model of the modern, industrial world that would be simple enough to be tractable yet complete enough to have some chance at capturing the essential factors of the real world outside the laboratory doors. For this model they chose the following to be the central families of variables: agricultural output, industrial production, human population, resource depletion, and pollution.

In October of 1972 the reading public was introduced to the results of the computer simulation created by this team of pioneering computer scientists at MIT. Opening the cover of The *Limits to Growth*, the 207 page mass market paperback, the publisher's blurb rang a historic wake-up call worth quoting in full:

"Will this be the world that your grandchildren will thank you for? A world where industrial production has shrunk to zero. Where population has suffered catastrophic decline. Where air, sea, and land are polluted beyond redemption. Where civilization is a distant memory. This is the world that the computer forecasts. What is even more alarming, the collapse will not come gradually, but with awesome suddenness, with no way of stopping it."

As we have discussed, it was in the 1970s that there was the first general recognition that the resource limits of our finite planet cannot sustain modern, petroleum based, industrialized civilization. The standard run simulation in *Limits to Growth* had the crunch time coming about forty years into the future, just about now. At the time it was published other news worthy events were seen as confirming how serious our plight was. In that decade there would be reports of collapsed fisheries, an oil embargo that produced gas lines and choked the economies of the overdeveloped world even while population pressures brought ghettos and slums of the inner city to the boiling point. Things have not improved since then; on the contrary and we have come a long way since the 70s. Homo Colossus stalks the very boundaries of peaking resources everywhere: fresh water, lithium, uranium, copper, platinum, grain harvest, oil, natural gas, phosphorus...

In 2008 Australian physicist Graham Turner while working for the Melbourne Sustainable Society Institute performed an updated comparison of the *Limits to Growth* with historical data. The results were published in *A Comparison of the Limits to Growth with 30 Years of Reality*.[14] The evidence from the last 30 years fits the business as usual scenario of the *Limits to Growth* modeling frighteningly well. In April, 2012 The Smithsonian magazine published an info-graphic based on Dr. Turner's work in an article entitled, "Looking back on the Limits to Growth: Forty years after the release of the ground-breaking study, were the concerns about overpopulation and the environment correct?"[15] The short answer seems to be yes. The data show that the trends that were predicted by the MIT team were real enough. The question that remains to be answered over the next few decades is whether or not the rest of the predicted changes will to pass, the ones that lay out how giant industrialization passes into the history books.

"...the myth of limitlessness had at last become obsolete."
Overshoot, William Catton

A Long Descent

What does collapse look like? This is an important question if we are to have a fighting chance of recognizing it. Look out your window. This is what it looks like.

There is a widespread, yet mistaken, notion that collapse comes on the wings of dramatic disasters whose importance is immediately noted by the entire world. If our most awesome high-tech civilization is going to collapse, surely it will not just slowly crumble away but will go out with all the glitz and glamor of a Hollywood block-buster, full of action, suspense and romance against a never ending backdrop of spectacle and miraculous special effects. The historical

Homo Colossus

reality of civilizations that fall is much more pedestrian, mundane in fact. We should understand that for those living through large historical changes, life, for the most part, just felt like life. Perhaps in "interesting" times there are more hardships, more scarcity, more violence, and there are certainly times and places where unmitigated disaster does lay its sorrowful hand on whole peoples at a time. None of this removes the continuity of day to day life on the whole. Even in a war zone the proportion of time spent under live fire is only a tiny fraction of a person's total experience.

The collapse of Homo Colossus will be punctuated by drama, no doubt, but in the main we can expect a few more broken things to just not get fixed, year after year, year in and year out. As resources become scarce it becomes no longer possible to maintain social and industrial complexity at previous levels, as Joseph Tainer's *The Collapse of Complex Societies* taught us. How does this play out in practice? We have been watching it take place all our lives. We take three steps down the complexity ladder and it allows us to take a small step back up. All the talking heads are hired to make such a big deal of that little step, it might even seem for a bit that the whole world was singing "Happy Days are Here Again" and surely flying cars are just around the corner for everyone. It took Rome roughly 200 years to fall, it could well be centuries before Homo Colossus breaths its last. It's likely to be a long, jagged journey towards whatever floor ultimately provides some lasting stability.

Much of this territory is covered by John Michael Greer in *The Long Descent: A Users Guide to the End of the Industrial Age*. The tonic his book provides serves as an invaluable corrective to the many Apocalyptic and Messianic currents of modern thought.

It is important to grasp that this concern we are involved in as we become mindful of ecology is a long term one. The work we are opening ourselves to participating in is not one which we ourselves will see much fruition in during our lifetimes. Nor, it must be said, will all the most important outcomes to affect our species be decided during our lives. We have our part to do but it is hubris to overestimate how important our generation is in the big scheme of things.

We inherited a lot of this mess, contributed all too much to it, and even cleaned up a little of it here and there, now and then. Much as our parents did. Much as our children will do. If you think there is going to be some final step that turns this whole thing around, or that there is going to be some crash landing so dramatic it wakes people up overnight, well, I am afraid you are only setting yourself up for a major disappointment. Nightmare dreams of currency crashes and other forms of the Zombie Apocalypse are easy ways to excuse ourselves from the hard work at hand.

That work is to begin to find the words and the courage to speak the truth among ourselves. This way when the larger events do occur, the larger steps down the stairway of the long descent such as currency crashes and war, we are emotionally, psychologically, and spiritually as prepared as we can hope to be. We can expect many a pied piper to arise in the land blaming our descent towards a less complex society on any number of factors from the rich to foreigners. Without an eye on ecology it will become increasingly difficult to maintain the narratives by which we weave meaning in our lives. With an eye on ecology we can spend those lives deeply engaged with a living, breathing preciousness that renews our dedication to fight for her health with every breath we take.

So where is all this leading? One of the postulates of this ecological point of view is that the chances of this whole fiasco becoming the end of our species full stop are slim to none. The Apocalypse has been canceled. What we want to avoid is blasting ourselves back to the Stone Age. Modern science unleashed the power of knowledge unlike anything seen before because it aligned with the molecular nature of the reality all about us. That we became enraptured, and eventually enslaved, by the machines this knowledge allowed us to fashion was the mistake. There is nothing in the nature of the molecular world to suggest humankind cannot make use of a type of technology that does not overwhelm the earth's natural systems. The water wheel need not hurt the stream. We need not forget what we have learned about disease and dentistry, or so many other things, even as we turn our backs on many other endeavors.

Homo Colossus

In the 1970s there was a movement towards critically examining the technology we had discovered over the ages. Lewis Mumford's *Myth of the Machine* provided nuanced yet trenchant criticisms of the megamachine. The idea began that with the right kind of design science it was possible to tame our tools, giving the alternative tech movement its name. The bicycles, windmills, and compost toilets that graced the pages of the *Whole Earth Catalog*, first published 1968, and *Mother Earth News*, started in 1970, were the first beginnings of something we can expect to characterize our children's lives for some time to come.

The point is, there really will be a tomorrow. We need to start acting like it.

5.
Seeing Blindness

Fighting Repression

It is our fate to live on the earth at a time at which our human societies are causing irreparable damage to life systems everywhere. Our societies treat this news just as any other, as if the end of an ecosystem were equivalent to the end of a marketing campaign. This is a mistake. It is a sign of confused thinking to fail to recognize the singular nature of the ecological crisis. It is incomparably more vital to our species than any of the other subjects that might be occupying the headlines in a particular news cycle. Given that this is the case, it then seems reasonable to conclude that before we can expect our societies to begin reacting rationally to our unsustainable lifestyles, some proportion of individuals will need to do so. When individuals fully integrate this ecological knowledge into their view of the world, the way they live in that world changes in response. Each person who learns to care about the suffering and irreversible destruction current social arrangements are causing the earth on such massive scales, also comes to their own unique way of expressing that concern. The emotions involved are too large not to be expressed. The steps Mindful Ecology suggests are, first, to educate oneself and then, second, use contemplation to integrate the implications of what you have learned. What further course of action you are moved to do in response to the ongoing crisis naturally develops from there. Though these are discouraging times, we need not live dark and depressing lives. When we learn to take meaningful action, actions that are meaningful for us as individuals, we find the antidote to despair.

Seeing Blindness

It is evident that the human mind is having a hard time grasping the true nature of our current ecological situation on this planet. For decades now ecological science has provided us with solid evidence that has left no question about the types of problems we are facing, how they will get worse, and what kinds of responses can be considered realistic. We have failed to act on any of this knowledge in a responsible manner. We still fail to give it the magnitude it deserves even as the evidence continues to confirm our worse case scenarios. In fact, our response has been so disproportional to the threat, that it stands as bleak, irrefutable evidence that most individuals and institutions are unable, so far at least, to come to grips with our ecological position. This is the point: the ecological crisis is caused by the lifestyles we have come to rely on in the industrialized world. On this the science is clear. It is not what we want to hear yet remains true just the same. There is no question among the scientists and engineers that whatever the future holds, it will undoubtedly entail the use of much less energy and many fewer goods. Today, to a rough approximation, our activities could not be sustained if we had four earths worth of resources at our disposal. This makes our existing arrangements profoundly unsustainable. It bears repeating that that which cannot be sustained, will not be. We can take one thing to the bank in our realistic planning for the future: the future is going to require that human lifestyles use less, less of just about everything.

To even mention this is not allowed in mainstream public discourse, beholden as it is to models of endless economic growth. This is not fate. We could refocus on what is needed. We could use our ingenuity to shrink our human footprint humanely, keeping the beauty of our communities and the dignity of our human relations intact through the changes of the coming decades.

For at least a generation or two all of us have known, even if only somewhat hazily in the backs of our minds, that the way we are living is not sustainable. We have never been a stupid species. We understood that if our technological exploitation of the planet continued to grow at its ever accelerating pace, sooner or later the bill would come due. This unwelcome knowledge has acted as a poison in our bodies and minds. It has also altered many of the things we do collectively as well, introducing an element of desperation to our commerce, religion, education, and politics. We see this in how the public voices insisting everything is A-OK have become shrill, as if facts could simply be shouted down, and we see it in the political systems answering primarily to those who benefit from current ecological exploitations becoming ever less responsive to the real needs of the earth's human animals, the people they purport to serve. Under these circumstances each one of us is presented with a choice: how conscious will we allow the findings of the ecologists to become in our own lives?

The last few generations who have lived with this knowledge have been characterized by people who used repression to defend against it. This is certainly understandable. Taking care of oneself and one's loved ones is what properly occupies most of an individual's time and energy. People of goodwill care about the larger issues threatening our societies but wonder what are they to do about them. They wonder why they should fill their thoughts with things that only make them feel hopeless. The repression approach does have a down side though. It allows our unacknowledged fears to fester in the dark. Such failure to attend to psychological hygiene can leave people vulnerable to exploitation both from within and without. There are many people and institutions in our psychologically sophisticated age who know how to manipulate people through their unacknowledged hopes and fears. The image makers, shysters, con men, and power hungry stand ever at the ready to exploit these vulnerabilities to obtain their own ends at other's expense.

An exploitation from without happens when a mind is hijacked by a totalitarian thought system. The victim becomes a true believer, convinced some person or institution has all the answers and the only true and good way to live. Such minds have been seduced by Pied Piper solutions packaged as fundamentalist faiths which come in all stripes: religious, technological, economic and political. An exploitation from within happens when the parts of a person just

Seeing Blindness

do not work well together and cause that person to perform actions they would never do if they were in their right mind. These exploitive phenomenon can be particularly troublesome in our world leaders. This is what repression is, this arena of psychological exploitations. Repression is a kind of bullying on the inside where scary and hurt parts of ourselves or our societies - and what those parts know - are kept silenced. It takes so much psychological energy to maintain the repression that it can drain the joy out of life. Repression also leaves us vulnerable to those who know how to push our buttons; they can become masters of our triggers.

The alternative to repression is, of course, to keep that which is being repressed conscious. The alternative, in other words, is to remain mindful of ecology. Something happens when a person is no longer kept spinning around by the parade of non-stop electronic distractions modern culture saturates us with. The mind left to its own devices turns inward. What one finds there, I suggest, is this call to change how we live in response to what we have learned about our exploitive human relationship with the earth. There is no magic here. Nothing special turns the sad, devastatingly sorrowful state of affairs into anything other than what it is. No great esoteric gnosis is waiting to free a few chosen from the calamities of a society rushing head long over the un-sustainability cliff. What Mindful Ecology is about is more basic, simple, rooted. It is seeking a compassionate response, one by which we may begin to heal the rift that has torn us from our felt connection with the earth and all her bounty. We have lost that connection which is our birthright; we could never do the things we do otherwise. I think that if we saw rightly the sacred nature of our planet, we would fight to the death to defend the health of the earth and the dignity of the creatures who, like ourselves, call it home.

Unless and until leaders speak of how the changes towards a more harmonious relationship with the natural world are going to become policy, they remain absent from the real challenges driving our history. It is not that what they do is inconsequential or even superfluous, far from it. By not recognizing the primacy of the ecological challenge and allowing that to order the rest of society's values, the political apparatus can only react to the ecological blow-back less than skillfully. Often, in fact, they are driven to pursue the very policies which are going to make things worse.

Eventually the balance between reality and the social organizations that has so far propped up this unsustainable system will tip. The suffering among the populations of the industrialized world will continue to take on third world characteristics. Part of the ecological crisis includes risks to global food security. Add to this the mass migrations of human populations from regions of drought and the basic outline of what to expect in the coming decades is clear enough. I expect the currently pervasive attitude that we will deal with the ecological crisis in a serious way sometime in the future to give way only when the triage activities of dealing with the blow back and fallout of our past activities dominates our daily lives. Parts of this process will take less than five, ten, or maybe twenty years to play out. This would be the test of the inflection points of the *Limits to Growth* model. Other parts of this process will likely take a century or two before the rubble finally stops bouncing. Rome, as they say, did not fall in a day. However these events play out in the country where you are, I am going to suggest that becoming mindful of ecology is the best preparation there is. It carries the highest likelihood of helping you analyze and respond to the changing situations skillfully.

Overcoming the repression of ecological knowledge and its true implications entails adding grief to your world view. But the tears are not without a balm, for they bring the gift of understanding. Cleaning out the repression is its own reward, for this is how we reclaim our full birthright as children of the earth, children of the stars. The question becomes, how are we to do this? Mindful Ecology suggests that a practice of meditative contemplation is a means people should consider to accomplish the restructuring of the mind and heart that is required.

Seeing Blindness
TV, Trance and the Collective Mind

I want to share a speculative model of why individuals and societies are failing to respond to the ecological crisis with a response rationally proportionate to the threat. The model is in the spirit of a psychiatric analysis of the developmental evolution of social imagery. It asserts the propagation of the belief systems that allow the destruction of earth's ecosystems to proceed apace depends on what amounts to an illness in the history of ideas. It is an illness in the collective conversation taking place between ourselves, and the other generations from which we came, and those which will follow long after we are gone. It is the working hypothesis of Mindful Ecology that this is the level on which our un-sustainability will ultimately need to be addressed. There may not be anything like the collective unconscious Carl Jung speculated about. There may not be a deep place within consciousness where all awareness is infinite, as mystics have reported since time out of mind. There may be no Aboriginal dream time. But, there can be no doubt, there is a collectively shared interface created where we express our individual subjectivity to one another. It is carried by our arts and letters. It is this mysterious arena we label the place of our cultural evolution, the public square, and quickly move on. Western thought is a bit uneasy about the exact status of its reality. This shared exchange of images and words bears the collective face of our ancestors. They speak in tongues with long histories. It is rather untenable to deny that there exists this shared mental atmosphere or, quite importantly, that each of us shares some responsibility for its tenor and content.

Nothing has had a more profound effect on this inter-generational exchange of ideas than the end of books as the primary medium of communication, and their replacement by the two dimensional projection of moving images. It is important to recognize each of us living in the over-developed world has been raised in an environment in which exceptionally advanced *influence technologies* targeted us from an early age. Most of these newer technologies shunned the printed word for the more direct manipulation of the mind's attention available with moving images and ceaseless radio chatter and songs. Nothing like this had ever been experienced by the human psyche previous to the development of electronics. Their effects have been profound, both socially and individually.

Contemplation teaches us to recognize the powerful role images play in the workings of the mind. This leads me to propose a model for how to think about our social inability to respond to the ecological reality our sciences teach us is real: society is suffering from a trance-like enchantment. We are behaving as though we are asleep or hypnotized. This is an extension of psychology's recognition of the importance of dream images, such as we see in the schools of Freud and Jung. This model argues that dynamic images are anchored in our physiological information processing systems, primarily the nervous system but also involving the endocrine and immune systems and whatever else works together to support our access and response to memories. All thoughts require this background of conceptual scaffolding, this memory field, as a generalized recollection of what is what, and what things mean. However, the dynamic images we are particularly concerned with in Mindful Ecology are not this run of the mill sort. The ones we are concerned with carry an emotional and physiological charge somewhat like the traumatic memories of those suffering PTSD. These cause our enchantment. These highly charged dynamic images make us uncomfortable because they are depressing and frightening when first encountered. But if we do not do the work needed to deal with them consciously, we begin to feel as though we have a minefield inside, one laced with explosive emotions holding us hostage. We are enslaved by what we dare not think. Cognitively, the result of such an unhappy condition is that our minds run along psychic grooves, the repetitive ruts, as it were, in the mud of lives lived in quiet desperation. Socially, we have an unresponsive, unquestioning faith in business as usual, the shiny thing we cannot take our eyes off of.

Seeing Blindness

These anchored emotional dynamics can be triggered by certain events that we have learned to associate with particular images, much as Pavlov's dog's learned to associate the ringing of a bell with food. Modeling what ails us as enchantment further assumes that these dynamic images can be triggered not only by events in our environment, but also by the thoughts we think in the privacy of our own subjectivity. There are some of these images, the model asserts, that act as guardians of orthodox authority. They are the inner policemen working for the business as usual crowd. One encounters them when sitting in contemplation. They really kick in when you try to think blasphemous thoughts: those not approved by the corporate image of the world carried in the commercially approved images of the mass media that today makes up so much of our shared memory field. Blasphemous thoughts, for example, such as these we are discussing about the un-sustainability of our industrialized ecological footprint.

Mindful Ecology argues our societies are suffering under totalitarian bewitchments on a massive scale. They do not come from a conspiracy cooked up by a Cabal. The bewitchments are simply the result of exploiting some of the human mind's psychological weaknesses for profit. Specifically, globalized growth economics, coupled with progress defined as industrialization's technological acceleration, has us so dazzled that we can hardly imagine a world organized any other way. We know from our ecological sciences that the healthy functioning of the earth's ecosystems on which we Homo Sapiens rely for our survival are all at risk from our way of life. Even knowing that this is now backed up by decades of data collection has not provided sufficient motivation for us as a society to do anything about it that is nearly proportionate to the threat. In my analysis this shows all the signs of an enchantment, a psychological bewitchment, an unrecognized slavery to a certain set of ideas to the detriment of the health of the patient. The cult of greed and fame has swallowed our culture and left little else. Provided a steady diet of images created by others, we can no longer imagine for ourselves that there might be any viable alternatives to our ecologically

100

suicidal business as usual arrangements. This is an enchantment. Like any cult member who has drunk the Kool-Aid, we cannot even imagine life outside these doctrinal walls. It was not easy to get the human mind to disregard the evidence of its senses in favor of abstractions like the GDP. The enchantment process is necessarily ongoing, sustained by the largest experiment in the mass induction of trance states that has been ever been staged: the glassy stare of eyeballs glued to screens all over the world, everyday.

External moving images have only been a part of the human psyche for roughly a century and a half. It is not idle to speculate that a future generation will look back on our grossly unskillful use of such powerful cognitive technology and wonder how we ever survived the fear and stress we put ourselves under. The widespread sex, violence, and torture images that have become so pervasive in our times are not even pretending to document real life. They have all been produced for the camera, certainly an evolutionary first. Life has always been R rated, and thank goodness the frequency of torture, witch hunts, public hangings, and all the rest of the horrors of our judicial past are not as common as they once were. Still, like the fairy tales of the *Brothers Grimm*, our tales are grim indeed. In a world where most people do not see beheadings or hangings when they step out their front door, we quite ritualistically gather around the modern camp fire light of our Silver Screens to watch our horrors in close-up, at just the right angle, with quick cuts from the editors designed to traumatically shock the viewing mind. I think they have succeeded. Our societies do not suffer the indignities of nature's cruel side, particularly when at times man acts lower than the beasts, with a stoic acceptance of the bad with the good. We go out of our way to create fake pains and horrors drawn in the most excruciating exaggerations our fevered minds can imagine.

Like shamans painting animals on the cave rocks of old, in our latest image manipulations all these tens of thousands of years later we still hope to gain some power over that which terrifies us. We have always been story tellers and here we are learning to tell them in a new and very powerful medium. Along the way it is not surprising that we have fallen into the traps storytellers are prone to: we came to believe our own bullshit. We are responding to the real as if it were another fictional story of our own creation, and responding to the fictional as if it were real. The difference between the news and the sit-com blurs in our decision making deliberations. Our society does not seem able to grasp the truth that it will not be able to re-write the ending of industrialized consumerism if it does not like how it is turning out. Reality looks poised to break the cardinal rule of the movie business: there is not going to be a happy Hollywood ending for the story of our times, the story of over-grown, hyper-developed, Homo Colossus.

The mass communications carried by these image manipulation mediums was quite consciously developed to have the most powerful influence on its consumers as possible. Influence is the name of the game in commercial advertising, and commercial advertising is the whole driver behind the rise of the mass media. Psychology has come to understand the way social pressures, status fears, and a host of other dynamics in the basements of our minds work. The instincts of the con man and the used car salesman have been supplemented with research, experiments, and reams of sociological data. Polling in politics is just the tip of the iceberg. Pavlov's dogs salivated at the ring of a bell because they were shocked and starved into imprinting associations that were nothing but coincidence. Something similar may have happened to us. Every time we use Dollars, Euro, Yen, or whatnot, the whole background context of television and movies comes along for the ride, and our environment suffers greatly for it. We can begin to get an answer to how and why by looking at a few of the more recent developments in the history of ideas.

Something changed after the World Wars with their industrialized mass murder in trenches, concentration camps and atomic bombs. Since that time industrialization's love affair with the machine has taken on a particularly satanic sheen, one we have worked hard to suppress. We were shocked by the betrayal of that which we had put our trust in. Mankind's dignity had been assaulted, very directly, through the means of industrialized R&D and factory floor precision, the very means we had placed our hopes in for creating a worldwide culture of peace and prosperity.

After the war, belief in progress through industrialized technology rebounded as European infrastructure was rebuilt with the help of the Marshal Plan. Though shaken and stunned, particularly in America, we were sure we could become master of our giant-sized industrial tools. Engineering as a profession formed around the increasingly important steam engines in the America of the 1880s and culminated in the largest, most complex public works ever attempted: putting a man on the moon in 1969. It looked as though the space age was the destiny of our species. We would harness atomic power, not let it destroy us, and electricity would become "too cheap to meter." This, too, did not work out. The computers used to assist us in our quest into space were industrialization's ultimate machines, thinking machines. Sadly, when we asked them to tell us about ecology, the relationship between organisms and their environments, what they had to teach us was very bad news indeed. By the early 1970s the *Limits to Growth* study had rather definitively put the planet on notice that our industrialized civilization was unsustainable. Our machines had delivered to us their final betrayal.

It was also in the 1970s that the advertising profession turned away from the widespread use of lies, or more graciously, misleading claims. Government regulation had begun to hold companies accountable for what they said in their advertising. Many an odd commercial aired at that time which explicitly contradicted the claims they had made previously. Listerine, for example, ran an ad on television created just to state the product would not help in the least in the fight against the common cold. The merchants of doubt[16] were pay-rolled by big tobacco in response to the new regulatory mood. Moneyed interests quickly understood that if facts were going to count, they could always pay to have a wall of lies and confusion thrown up around them. That these same fine specimens of our species are busy discounting climate change is so well known it has become a truism. This, however, is not the only development within the advertising profession with which we should be concerned. The ultimate outcome of regulatory attempts was to have emotional messaging replace the communication of information. Behavioral manipulation through emotional appeals proved to be an astonishingly effective means for dealing with the new regulations around lying to sell your product.

Half a century after this flood of emotional manipulation was unleashed upon the world, we find ourselves socially dysfunctional on the macro scale in almost every way. Nor, I might add, are most people very happy and fundamentally content with their lives. Turning this around will necessarily involve working with this emotional conditioning, particularly at the semi-conscious levels where it is most powerful. It will also involve learning to live, day in and day out, with the cognitive burden of what we know about our dire situation, the knowledge we try to repress at all costs. Doing these things involves working directly with the mind-body in which they are anchored. In the age old tradition of meditative contemplation, we are given the opportunity to encounter these things directly.

Meditative Contemplation

Contemplation, as I am suggesting it, might take an hour or more. This is not shallow thinking where we note a few facts and quickly move on. This is slow and careful absorption of information on every level of our being within an overall context of mental equipoise and bodily homeostasis. It involves mental images but is not an exercise in visualization. Instead the images are used as poetic metaphors through which emotions express themselves. At times during the sessions, this mix of knowledge and metaphor sparks the intuitive perception within us to deliver lessons only our own internal milieu can teach. First we calm our minds, touch the silence within. In this safe internal container we then allow our contemplation to proceed. If we find our minds starting to race or lose focus, we gently return to the silence, trusting patiently in the process more than any particular content. That, in a nutshell, is the type of contemplation practice Mindful Ecology is discussing.

Awareness of the ecological crisis demands a response. Adopting a regular practice of contemplation is one way of expressing a dedication to the healing of the earth. These contemplations aim to put the practitioner directly in touch with their own wisdom, confident that, ultimately, no one can tell you what you should do in response to the needs of our time. This is between you and the earth alone, your body and its journey from womb to tomb. Everything you will ever know and everything you will ever do will unfold here, on the surface of this earth, under the ever watchful stars. The self that is the sum of our whole lives is greater than the self we are moment by moment, yet, it is not different from us in those moments either. In mediation we touch both. In the hustle and bustle of the modern world it is easy to get so wrapped up in social concerns, that we forget our more fundamental roots in the elemental, molecular world and how our flesh shares a biological membership that extends to the rest of our DNA family, all the living things with which we share our rare and precious earth. A lifestyle that includes a contemplative practice seeks to redress this imbalance between what we seem to be on the outside and what we seem to be on the inside. It does this by making a space in which the other dimensions of our being, besides the socially oriented ego, are able to express themselves. In our practice we remember our existential situation every day and train to say 'yes' and 'thank you' for the opportunity that is life.

This is not an easy path. Part of the power of contemplation comes from reminding ourselves that we are not permanent phenomenon, that one day we too will die. Another part of the power of contemplation comes from accepting just how small a part your life will play in the big scheme of things. Contemplative work necessarily involves working with our deepest hopes and fears around these things. The mind has a wild imagination and it is not always unconfused about what is real and what is a metaphor. The greatest fears of the mind are those it creates for itself out of its own imaginings.

Seeing Blindness

The startling finding is that training the mind to recognize those things it should really fear is the shortest path towards over coming the imaginative fears it has burdened itself with unnecessarily. It might seem that reminding ourselves regularly of our mortal limitations would be a recipe for pessimisms of the worse sort, but in practice we find just the opposite. Only such reminders keep our awareness fresh and clear. We need them to stay awake to just how precious each day really is.

Adopting the contemplative lifestyle involves a type of voluntary simplicity. As contemplation deepens we naturally remove clutter, both mental and physical. Simple pleasures resonate with the cultivation of gratitude, and gratitude teaches us contentment. Ultimately, contemplation offers us a way to feel at home on the earth, in our skins, and living in our time. A sense of place develops, soothing deeply held fears. Ecology, the study of the home, bears its fruit.

Our bodies are responding to the ecological crisis, we need only learn to listen. We know in our bones, as it were, how the work before us needs to be accomplished. Each person's path is as unique as their DNA sequence. To walk it well involves getting in touch with the inner authority your heart can trust, an aspect of our psychological experience that offers guidance, encouragement and teachings that speak most deeply to you as an individual. This is a place deeper than ego, larger than ego. It is what we can trust to see us through. It is worthy of our faith. It has been called by numerous names: the inner guru, the Lord, the capital S Self, the Holy Spirit, the Holy Guardian Angel, and many others. The point is, that once a person has been made right in their heart, they can trust that their inspirations and intuitions will never lead them too far astray from what is real. When we have learned to lead with the heart we take our seat, accepting advice from other people but never allowing them to judge us, shame us, or otherwise evaluate our ultimate worth. Only that which knows all things about me can do that, that mystery from which I am made real.

Getting in touch with this inner authority gives us the skills we need to navigate the world of external authorities. The ecological crisis has been brought about, however inadvertently, from the practices of business as usual which continues to have no shortage of spokesmen. It has never been easy to trust one's own lights when on every side we are surrounded by people and institutions insisting that they alone have the truth and that we must give them our alliance. It is the individual, not any group, that carries the creativity that will see us through the end of business as usual and on to whatever lies beyond. The ability to remain confident in one's own inner authority becomes crucial for those who are called to witness to the ecological devastation of our times. It is not easy to stand against the tribe when the tribe is making an error.

What follows is a collection of points to provide some shared context between us as we continue to explore the role of contemplation. First and foremost is the fact that there is no matter in all the known universe more complexly structured than the human brain. This is our touchstone and it implies a number of things.

* We understand we need hours in the gym to tone our bodies. It is reasonable to expect a similar time commitment to the art of toning our minds.

* We should approach the exploration of our mind with respect. Often in the mind not much seems to be going on, when in fact the limit of what the conscious mind will be able to endure is churning just below the surface.

* Though we may have a subject we wish to consider in a contemplative session we do not force things; we sit with it, watching, letting things happen, guiding it with all our skill but ever ready to return to the quiet when we need to ground and comfort our innermost being.

Seeing Blindness

* The Western insult of meditation derides it as "navel gazing." The insult is born of the West's fear of subjectivity, yet notice it includes a decent intuition about DNA if taken literally, or about the inner work with parental "parts" when taken metaphorically. The traditions of silent prayer in the West honor meditation as deeply as those of the East.

* It is possible to use anything sacred to perform what is called a spiritual bypass. The sacred should act as a platform for accessing the existentially profound mystery around the reality of your existing. Once you are right in your heart you can trust the intuitions of the inner guru to lead in a way conducive to overall maturity. Putting the same point another way, you can use meditation to drive yourself even crazier than you already are. Watch for using it to feel you are more special than anyone else, particularly those who do not meditate. When this comes up remind yourself it is, among other things, a healing practice you are involved in and that you are as deeply involved as you are because you need that much healing. Watch for feeling that you are acquiring a special knowledge never known to anyone, a gnosis revealed to you alone or to your group alone. You can be fundamentalist about the practice and lose the whole point.

* Contemplation and meditation is one of the most ancient arts. The traditions of deep introspection are many and varied, yet we can be assured that every person who has ever sat down to meditate encountered the same raw material of mind that you will encounter.

One final, most important point: you must be the one to decide if you will explore, adopt and perhaps remain committed for life to a meditative practice. Though this book is among countless others in extolling the ways of meditation, do not take my word for it. It is your choice to adopt such a practice or not. If you do proceed to do the work, and one cannot expect results without doing the work, retain the same attitude. Every step of the way test the work and its results to determine if what is happening rings true, real and good to your best lights.

As mentioned, there are times when working with meditation that it seems as if nothing is happening. We can be tempted at such times to do more radical things to try and force the experiences into some mold or another of what we expect or want to have happen. The wiser response is usually to stay the course. There is a natural ebb and flow to these things. Most often the tide will turn again and the practices you have been doing will begin to look like maybe they have been a bit too much. A touch of fear for your sanity, well-being, or character may arise. This too is just part of the ebb and flow. The advice handed on to me for these times, up or down, was to stick with the form of the practice and carry on.

This is the nature of a mature practice. The images and emotions that accompany sessions can bore us, even though they are powerful. Patience is called for, and a gentle touch. Just as we often do not appreciate what we have until it is gone, we often do not realize the power of a practice until we go a touch too far. There is no success like excess.

What does too far mean in this context? Basically any temptations not to take your body with you are to be resisted. This means always respecting the physical limitations related to proper nutrition, exercise, non-poisoning and such. The body-mind can only take so much 'light' at a time. In fact, the body is more comfortable with soft, warm glows that carry an animal nurturing than the kind of internal light shows that the mind can conjure. Push the body-mind too far and you will exhaust it. Instead of increasing calm and compassion, you will experience an increase in irritation, desperation and fear. Let go, you are grasping too tightly.

With experience each person comes to understand how to move close to the edge where the meditation is deep and powerful, yet the dangers are minimized.

Seeing Blindness

The other danger someone just starting to train in these arts should be aware of is the mistaken idea that just because one is now meditating all the rest of life should become what you always dreamed it could be. Another form of this temptation is to think that because one is now a practitioner you can neglect the mundane responsibilities of everyday life. The teaching here is unambiguous:

Before enlightenment - chop wood and carry water
After enlightenment - chop wood and carry water

Not to give away the show before you have had a chance to participate, but it is worth pointing out the path seems to lead to a more thorough commitment to the elements of our existence - the fire of our warm body and the volcanoes, the water that makes up 60% of our cool body and the clear running mountain streams, the air in our lungs ceaselessly breathing in and out in an intimate exchange with the vegetative chlorophyll, the earth of our bones and the mountain rocks filled with minerals, and finally, it asks of us a commitment to the emptiness, the space within our organs, our veins, our synapses, and the external space of infinite extension in which we move.

It is best if we can practice for its own sake. It is said to be the attitude with the most wisdom, perhaps because it recognizes the existential reality of how exquisitely precious these opportunities are. Of course, much of the time we will be dull to that existential reality, cocooned to one degree or another among our conceptual abstractions and light years away from the perceptions of the moment. At times like this, having a model of what it is meditation accomplishes can be helpful. I will provide one such model arranged around neurological imprinting by and by, but the point here is that we should not approach meditation only as something we can get something from. We should also nurture the attitude that in meditating we are giving something back, letting something go; we are saying yes and thank you to a gift that has already been given.

At Home in the Unknown

There are no guarantees that by adopting a meditative practice we will find the power and insight to turn around the ecosystem destruction our generation is participating in. It might not even bring us much relief from the anxiety or the deeper sadness involved in knowing how bad things are. In my experience it will but your mileage might differ. What the adoption of a meditative practice will do is provide you with an opportunity to explore for yourself the mystery at the heart of consciousness. When we sit to meditate we join a whole host of fellow explorers of the inner world of the human mind that have gone before us. It would be a real shame to live and die without ever really nurturing the awareness of the mysterious vastness of the universe, the existential oddity that anything exists at all, the wonder of personality, love, compassion, and friendship. Busy as we are making money and dealing with social conventions, there is a danger that life will slip through our fingers before we ever really had a chance to come to terms with it, really had a chance to explore its deepest reaches.

The future is uncertain. The ecological crisis just adds to the weight of unknowing. We are numb to the full contingency that accompanies our journey through life and our planets journey through the solar system (not to mention the solar system's journey through the galaxy and this galaxy's journey through the galactic cluster). Though change is the only constant, our nervous systems provide us with an image of environments that seem static for the most part. Our awareness is structured by our perceptive apparatus so that it seems to us events unfold against a static background. These structural components filter the differences and juggle our ceaselessly changing biological states to highlight the similarities and consistencies. The abstractions we find so easy to use, even simple ones like our labels for say a chair17 encourage us to pay attention to generalizations at the expense of particulars. Too enraptured with

111

our own generalizations, we find this seemingly static consistency all around us, as if it were a characteristic of objects themselves and the environments in which they are encountered. The danger is that these generalizations can blind us to the particular nature of the actual object: there exists nowhere a 'chair' but there is this chair and that chair. Nor is the chair that exists in this moment the exact same chair that will exist in the next. And what is true of chairs is true of everything else, ourselves included. This is simply the real and true nature of the molecular world.

We fear this vision of the radically momentary leaves us completely without ground: if everything is changing and transforming endlessly, where can we make a stand from which we can protect ourselves and our loved ones? The teachings here are related to so-called selflessness. Our individual experience of conscious awareness is that of our own emergent consistency, the continual out-flowing of self. To be human is never to know a moment without a body and never to know a moment without a mind, however much it may seem at times that this is not the case. Life is this out-flowing. It is the pure gift of a stable, structured emergence within the ever changing hum and buzz of the molecular world. In other words, the only way to make it through the ring of fire that is our fear is to set aside the point of view that is exclusively concerned with our personal survival, or what often amounts to the same thing, our concern with our social role, where we fit in the tribe. We learn to set aside the fear in many ways. One way is when we tune into the inner witness, that part of consciousness that is able to simply observe without getting caught up in what is being apprehended. Another is to learn to love that which is other than oneself for its own sake, to be happy just knowing it exists. There are dolphins jumping and wolves howling right now, it is good to remember that. I have found contemplating a pure mountain stream can often provide a touch-stone for practice sessions in which difficult emotional material is being processed. Just knowing the pure water cycle is carrying on throughout my watershed somehow allows the pain of human folly to recede a bit into the background, taking its proper place within my grasp of the big scheme of things.

Each precious day we awake offers an opportunity for the universe to be awake to itself through us. No one really knows just why things are this way or what this ultimately means - if it means anything at all. The human mind as participant cannot step aside and look at the whole as would be required to answer such questions. Kant and Schopenhauer demonstrated conclusively that thought not dependent on our human categories will remain forever outside our grasp. Even the self we postulate as the sum of a lifetime (or a million lifetimes) remains beyond what the self of the moment can ever fully know. We can, however, become comfortable with the unknown, as if it were our natural home, as if it were our own most intimate being. Which it is.

We relax. We learn to trust the ceaseless coming forth from emptiness. *With the courage to accept that we really do know what we know, the unknown no longer frightens. It titillates.* We now act instead of react as we encounter, again today, an ever alluring invitation to use our limited understanding of the relative, molecular world to do something meaningful. Learning how we ourselves can live a meaningful life, right in the midst of our difficult times, is what we are seeking by remaining mindful of the full spectrum of our ecological relationships.

6.
Meditation Advice

Forwarning

In the next chapter we are going to talk about a few of the ways in which a meditative practice can be formed around ecologically inspired contemplations. Buyers beware. These types of tips and advice can, I sincerely believe, help us when we share what we have learned about being on this difficult path. They can also be misused as excuses to avoid personal responsibility for how we treat our minds and bodies. To avoid that sort of silliness, here in this chapter are what I consider a few basic points to act as a sort of contract with my readers.

1st point - inner guru. I use this term to refer to the part of our mind that is wise beyond our wildest dreams. Some would call it the Holy Spirit. It offers only compassion and wisdom as one and the same thing because he/she/it understands you better than you understand yourself. To truly understand is to have compassion. Developing a sense of this inner guru is the point of so-called spiritual exercises. To guide us finding it, it is taught do not believe in any voice that talks to you with anger, hate or rage about your mistakes and worthlessness. Those voices seek your self-destruction. Most of them seem to come from abused parts of ourselves we have repressed. They hide valuable things for us once their unrealistically heroic burdens have been released by the power of the inner guru's compassion. In meditation there is a type of shamanic healing involved, or that of a good shepherd willing to find and succor the lost and outcast.

2nd point - authority. You cannot disavow that you are ultimately responsible for anything that happens if you choose to do any of the exercises I mention. Many self-help books come off as if they had all the answers and you just need to accept their program hook, line and sinker. Those are not all that different than fundamentalist religion in my mind. I want to scream "That is not the case," the classic response in Tibetan monastic debate. It is not the case that any of these sources have all the answers or even just those that will apply for you in your search for a meaningful life. In fact, in either the self-help or religious claims, even if they did have all the answers, they cannot have all the answers for you anyway since your truth can only be claimed by you, not force fed.

3rd point - This book is the result of my long struggle not to go crazy with grief for our callous ways and selfish blindness. I do not consider myself a paradigm example of the healing that might be possible through mindfulness practices. At times I am just barely hanging on, but I no longer hide from what I know to be true. For a long time I just drank too much. It helped me in the oscillations between some happy days where I suppressed my knowledge and the scary and depressing days when I could not hide from myself the terrifying world my knowledge had revealed. Contemplation eventually provided a container for what I had learned. It is hard to know that our home planet is being killed one animal at a time, one coral reef at a time, one tar sand pollution-fest at a time. Our bodies are reacting to it, absorbing the toxicity, and it is coming out our mouths and actions as so much violence inspiring hate and greed. Cut off the knowledge of the eco-crisis from your awareness by denial, repression or distraction and you cut off the energy of being fully alive in our time. To be a caring human being is the point of being a human being. To be a caring human being today entails dealing with your reaction to the destruction of ecosystems our lifestyles are causing. And who knows but that there just might be a benefit for some. Alternatively, if you bury your head in the sand, do not be surprised if the scythe of history chops it off.

The Point: We engage in learning and sitting with ecological facts so that, as they inform our hearts and minds, our daily actions will become more skillful. The human relationship with the earth and her resources is sick. The point of meditation is not meditation, which would be silly. It is to become the most skillful healer of this soul-crucifying disruption we can possibly be. By developing integrity and clarity around our convictions, our actions naturally become more skillful. Each of us, in our own ways, need to learn how to walk lighter on the earth. But first we must get right in our hearts, give ourselves time to mourn, remove the toxin that eats away at our proper human dignity. We need to school ourselves in exactly those virtues our culture derides: patience and contentment. For then, I believe, our work becomes not only one of mourning, but also a source of great joy.

A Word about Cults

Introspection in the Western world has a rather suspect reputation. There are some good reasons for this. A number of psychologists have had to deal with the fallout from cults that use altered states to bewitch their members. This is a very real phenomenon. Totalitarian thought systems, those which claim to have the final answer to the puzzle of existence and demand unquestioning allegiance, are found scattered throughout religion and politics. For the minds caught in the cult's self-referential double-binds, the absolute truth of even the most bizarre dogmas seems unassailable. Of course, to any outsider all the minutia and esoterica that accompanies the true believer's religion or politics seems downright crazy. This is the enchantment of the mind, a bewitching that is all too real. Most of the public recognizes that groups like the Moonies and Hare Krishnas have an *undue influence* over their members. Somewhere a line is crossed between freely chosen group participation and that which is psychologically coerced.

In Mindful Ecology we are interested in removing consumerism's anchors, to use a term popularized by Neuro Linguistic Programming (NLP). Many of the same persuasion techniques found in groups practicing undue influence are recognizable in mass communications as well.

In *Under the Influence: The Destructive Effects of Group Dynamics* John D. Goldhammer brings a Jungian point of view to the discussion of undue influence. The work is highly recommended. "Spiritual ecstasy," he writes, "as one extreme of consciousness, is a perfumed trap, a sweet poison in the soul, an infection of light possession - all air with no feet on the ground. Imprisoned in a childlike world, individuals caught by this ecstasy-of-light experience are unable to deal realistically with life's existential issues and conflicts... We humans have a built in propensity to leave the earth, to escape the weighty side of living - a sort of innate suicidal impulse. We create conditions for suicide by not living our own lives. Thus, groups that emphasize spiritual escapism reinforce suicidal impulses... Being caught in a religious cult was a significant death experience for me. It enabled me to substantially differentiate my thinking process from the collective. As James Hillman maintains, 'the death experience is needed to separate from the collective flow of life and to discover individuality.'"[18]

Robert J. Lifton's study of thought reform, *Thought Reform and the Psychology of Totalism: A Study of "Brainwashing" in China* has this to say, "The most basic feature of the thought reform environment, the psychological current upon which all else depends, is the control of human communication. Through this milieu control the totalist environment seeks to establish domain over not only the individual's communication with the outside (all that he sees and hears, reads and writes, experiences and expresses), but also - in its penetration of his inner life - over what we may speak of as his communication with himself."[19]

Some psychological sophistication is required if we are to understand a bit about what it is we find when we dare to look within. This is not to say psychology has the last word, but it is a word it is not safe to ignore. Understanding how the double-bind leads the mind to run endlessly around unsolvable dilemmas helps us all stay on the lookout for stinking thinking.[20] Being aware of the hold cognitive dissonance can have over our ability to reason clearly provides some honest humility that just might protect us from the most grievous errors which accompany those who are certain of their certainty, true believers of whichever fanatic stripe.

As JoAnna Macy teaches, we are traumatized by the ecological crisis when we are forced to bear it alone. We are not meant to bear this burden in isolation from one another. We need to find people we can share our fears and frustrations, as well as our hopes and laughter with. These few words of concern about the unscrupulous are not meant to discourage people from joining groups, just the opposite. They are offered as a way to safely navigate one's way.

Pointers Pointing

"Relax into a cheerful mind."
    Dzochen Ponlup Rinpoche

When you woke up this morning did you take a moment to appreciate your position on a blue-green globe sheathed in life hurling through space? Before putting on your work-a-day persona, did you take a moment to appreciate, aka offer thanks for, the humming cellular symphony of the body-mind in which your awareness just woke up? Did you notice, even if it was only a nanosecond, the mind's gyroscopic orientation as it put both of these inputs together and sensed being held in the close grasp of all that is?

I was taught to meditate at a young age, rather unusual for a white middle class American boy. Now in my 50s I have found a couple of things about it that I would like to share, things I wish someone had told me earlier on.

Meditation is presented in the West these days as all sweetness and light. It is not. You are exploring the mind, many parts of which have very primeval roots. Patience and a comfortable warm glow will get you farther than light shows. Don't get carried away by images of sunshine on glowing yoga mats in the high class gyms of the beautiful people. Trust the inner intelligence and respect the underworld; an unpleasant dream here and there is OK but too frequent or too severe and you need to back off a bit, let things breathe. There are seasons to do by not doing. Depending on your individual character you might need to back off for months or even years to digest some of what you learn when you begin to learn about ecology.

One of the most important things this book wants to share is a correction or refinement of what many people think mediation is all about. Mediation is often presented as an exercise in shutting off thoughts but that is all-together too simplistic. Slowing thoughts, shifting awareness deeper than thoughts, entering awareness of perceptions and senses - all these might be ways to say what it is actually like to be in a profound meditative state. Expecting anything more is setting yourself up for an exercise in frustration and failure (which can be a good teacher too).

Another misunderstanding concerns the role of quieting the mind. It is not an end in itself. The calm mind is used as a support for a type of slow, deliberate, careful thinking often called contemplation. The quiet and the contemplation are traded off each other. First quiet is cultivated as the stage on which the subject of our contemplation will be held in a soft focus. Bringing up our subject we just seek to dwell with it awhile and observe the reactions of both body and mind. When the mind becomes too discursive or conceptual we return to cultivating the calm, letting the contemplation go. Once quiet is established again the contemplation continues.

In contemplation we are opening ourselves up to the intelligence we embody that is larger than our conceptual mind through using the conceptual mind in an unusual way. When the subject of our contemplations arises in the calm mind, our organism is able to process and integrate its ramifications more thoroughly than normal thinking allows.

There are numerous benefits from developing the knack of using our minds this way. It is a method for taking what we know to be real and true intellectually and teaching the rest of our habit-driven emotional, intuitive and sensual being about it. The central motifs traditionally are contemplations of the infinite nature of the universe and the unfathomable reaches of deep time. These are ideas the conceptual mind can toss about rather glibly, yet fail entirely to grant one's everyday life the unassailable sense of dignity and gratitude they invoke when their reality is "felt" as well as understood intellectually. Another family of subjects traditionally explored using these techniques are those that trigger profound emotional reactions. Thoughts of our own mortality, the pain and sorrow our lives and the lives of our loved one's have known, injustice and man's inhumanity to man; all are traditional subjects for contemplation. The context of the quite mind allows us to explore our fears, angers and lusts without getting caught up in our typical stimulus-response reactions that are conditioned into our nervous systems by the intensity of the experiences involved. When the reactions disturb our equanimity too much, we simply return to the quiet and let go of the contemplation.

The contemplation of these traditional subjects are said to slowly transform a person's life if they are allowed to, bringing wisdom and peace of mind from which compassion for all beings naturally flows. Compassion is the other touchstone of a meditative practice. Just as we learn to return to the calm and quite mind as the inner landscape from which we venture forth into the explorations of our contemplations, so the motivation for why we are doing what we are doing is never far from the surface of our minds. The golden teaching is that we engage the mind and pursue wisdom for the sake of others. Accepting how much others have cared for us and how much we care for others is the ground of integrity on which all our practices and studies should take place. We have been loved and we have loved. This is what is most real about us, the lifeline to the heart of existence. This is how we stay grounded in the reality of interdependent existence.

One of the time tested means by which our compassionate nature can be nourished is through devotion. Devotion is the heart of all religions from the point of view of the mystic. The mystic values a person's individual interaction with spiritual matters above all else,

Meditation Advice

seeing here the source and goal of all teachings. In meditation this devotion is offered to transcendence within an imminent sacredness. The mindfulness of the breath, the beating of the heart, the flow of information through senses and thoughts are all a gathering together of an awareness of the immediacy of our existing right here and right now. This is the only place we will ever encounter that which is, for us, our heart's deepest desire and most sacred dream. To meditate well is to learn to become comfortable with the nakedness of this raw impulse that exposes us to disappointment and ridicule, and yet holds the key to our happiness: the courage to honestly say yes and thank you to the mystery.

In my way of understanding, you should not be too fixed in what you offer your devotion too. I talk about the inner guru as a way to indicate that there is fountain of intelligence and wisdom as close to us as our pulse, always available to us to turn to. What form it takes should be allowed to change and even remain nebulous and ambiguous at times. Otherwise we risk projecting our ego on the stars and worshiping its hopes and fears. Our task is to develop our devotional sense, this heart felt yes and thank you. The old words of advice in the West were to "enflame oneself in prayer." We learn to say it even to those parts of our existence which may make no sense to our sense of right and wrong. Gratitude recognizes what interdependence implies: that for anything to be the way it is, all things need to be the way they are. There can be no flower without weeds; the kiss you share could not come to pass without the collision of galaxies far overhead.

So these are what my experience as taught me are the main elements of a meditative practice that leads to healing and that can be sustained day after day, year after year. It would be wrong to think of such a practice as a duty or a burden. It is actually a joy and pleasure, an honest little-boy curiosity is never too far from the energy mix that keeps me going. This is the final bit, the attitude to take. Refuse to let life slip by unexamined. When we sit we are saying we are going to handle for ourselves the materials that make up the mysteries of our being, so that we will not arrive at our death only to discover we never really lived. Meditation is all about being open to the childlike wonder which marvels at the beauty and miraculousness of life, the universe and the mind manifest within them both.

I would like to close this section on advice with a collection of teachings and tips that might prove useful.

Compassion for self and others is the path, and every step on the path. Wisdom develops as our acts of compassion become more effective. Compassion for yourself is often the hardest. It entails seeing your mortal life as one equal in value to any others; not less so, not more so. Equal.

Hold very, very still at times of deepest absorption. Do not develop the habit of rocking back and forth.

Calm abiding is the Buddhist name for the quiet part of meditation. It is likened to a vase protecting a flame from the winds of scattered mental discursion and focusing it so that it becomes brighter.

Take time to feel the world instead of always trying to figure it out or make it do your bidding.

Passion - what you wake up in the morning wanting to do. Your contemplative session should be pursued with passion. It is useless when it is just a chore. Each time you sit expect complete, perfect enlightenment.

Gnothi Seauton, often translated as 'Know Thyself' is actually '[human] know thy place.'

Don't talk about results or cling to any in particular. Each session is kept fresh that way.

Working with the eyes:
Meditate with them slightly, comfortably open. These sessions seek to reawaken us to the real world, not getting lost in internal revelries. Experiment with looking down about three feet ahead, straight ahead to the horizon, and up to the sky. Lower the gaze to slow thoughts, lift the gaze to fight stupor. Use a wide field of view to enlarge the sense to its full awareness or turn awareness of it

Meditation Advice

down a notch or two and provide inner space for contemplation. The exception is when the session is going to be devoted to concentration on an object, in which case they are made to rest gently on the object of your absorption.

After establishing ourselves in calm abiding we begin contemplation. In contemplating we say 'this is what I know.' It is difficult because the result is to simply highlight how much we do not know. This is the don't-know mind, beginner's mind of Zen and is to be encouraged. Being awake is not having all the answers, being awake is a direct confrontation with our puzzlement closer to the skeptics way than that of the true believer.

When we beat ourselves up we are in the grip of dualistic abstractions. One part of the mind attacking another part is not the way to peace. Not listening to that voice, challenging it when it arises, one begins to make friends with the mind, the mind begins to trust you a little.. You will sometimes find the mind referred to in Eastern teachings as the monkey mind. The first meditations can be like grabbing a monkey - scratch, bite and piss seem to be all you get. People can think meditation is making it worse, when really what has been going on all along is just being noticed clearly for the first time.

I recommend keeping a practice journal, at least for awhile, a few years. Simple entries are fine - date and time spent and type of practice. It provides a place to record highlights and other experiences. The trick is not getting insights; the trick is learning to integrate them into the rest of your understanding. This integration can be further encouraged when contemplative experience is reprocessed through the language centers of the brain as they are written down.

Assimilating Sad News

What do you do when you encounter a piece of horrifying news? You are going about your day, dealing with this and that when through the radio or from a web page click or in a comment from a friend you are suddenly struck by something that shakes you to your core. These moments are not typically full of fireworks, just very deep and powerful feelings are invoked, and your heart-mind gasps. Hurt before the psychological walls of your character armor could be erected, this arrow with its poisoned barb has found its way into your inner sanctum.

To process these shocks requires that we respect them and ourselves enough to just be with what we sense for a while and let it naturally evolve to wherever it is going to go. As most of us are not free to fully process these shocks at the time they happen, we are all very good at carrying on as if nothing had happened. We suppress the full assimilation of the news into our being for another time. But, since we are rarely educated in how to sit with these parts of life or why it is important, many people never get around to completing the assimilation work for the first shock before the next one comes along. And the next and the next. Eventually a generalized numbness to evil and pain develops like a thick scar tissue over our hearts, over our raw feelings. In this sad state even our dreams can only bring us a little truth since we are unable to bear much else.

The alternative is to remain mindful throughout the day. Not a big deal really, just being on the lookout for how our body-mind is reacting to the news and events of the day. With this watchfulness we become more skilled at recognizing the psychological attack on our well-being when it happens. We catch it before the auto-numbness swallows it and we go back to pretending everything is ok. Now we have the chance to bring that which impressed itself on us so deeply to our meditation.

Meditation Advice

The mediation of absorbing sad news, as I have experienced it, goes somewhat like this:

Once grounded in the calm we bring our image or sense of the inner guru to mind and once we are in the presence of that which accepts all we are with wisdom and compassion, we allow those powerful reactions to arise again, here in a safer place. The trick is not to add energy to them artificially but also not to cut off any of the real energy it might carry. Visual images and non-conceptual feeling-sense alternate as the body deals with the fear, pain, despair, horror, sadness, and shock. If thoughts start to spin into extremes, we drop everything and return to the calm. What we are conscious of is how this bodily processing is experienced in the mind. If the whole thing stays on the level of right and wrong that is ok, but it means we have not penetrated very far into the felt reaction, choosing instead to remain within the more conceptual realm. This really is ok as that might be all we are able to deal with at this time. Compassion does not push, still, it is also without fear so it will not settle for less than the full assimilation you are capable of.

Eventually something will move, get un-stuck in a way that often feels as if there has been a thawing of a frozen knot of sorts. Basically we sit with the hurt, just being a witness to the futility of the cosmos until our state changes to once again allow in a little bit of warmth or light. This is not unlike reintegrating a traumatic memory, sewing it back into the rest of life, which brings healing to those suffering PTSD. The state change might be a re-dedication of energy to resist and fight that which is causing the hurt. It might be a perspective of deep time in which the current pattern no longer seems to dominate all the meaning of the universe. It might be an interdependence insight in which that which causes the hurt is recognized as a necessary part of the current time and place, though without any inherent existence that would insist on it perpetuating itself forever into the future. It might be some other insight but they all become the means by which the solidity of the problem, which felt like it had life trapped, grows porous. We learn to recognize the space around it in which it is workable.

What we struggle against is the set of lies that say there is no power in the force for good, only in the force for evil. When it seems that way, when it seems the bad guys always win and have all the power, we have simply overlooked reality. The acts of simple kindness and love far outweigh the acts of terror, this is simply the truth. Those acts of loving kindness have not ceased their radiant witness and celebration of this precious fleetingness that is our lives. The untrained ego is not adept at noticing these acts of kindness, they do not seem worthy of its attention. It feels entitled to them so they tend to escape notice, disguised as they are in their cloak of ordinariness. Mindfulness notices them, pauses with them a moment to savor them. Whether it is an act of kindness you were given the opportunity to do, or one someone extended to you, or even one serendipitously seen between strangers. A good example of what I am talking about takes place at the start of the 2003 film Love Actually. It is the scene that gives the film its title. The camera shows a busy holiday airport with friends and family, lovers and first time acquaintances, children and parents all greeting one another as they arrive from their many different destinations. The voice over softly speaks the truth:

"General opinion is starting to make out that we live in a world of hatred and greed, but I don't see that. Seems to me that love is everywhere. Often it is not particularly dignified or news worthy, but it's always there."

There are very dark events unfolding in our world today. They are all around us, we cannot escape them. If we close our eyes, as the poets say, we still absorb them through our skin. Allow a place for them in your view of the world, but learn their place.

Hold the eternal falling in love between us sacred. Think like a mountain.

Authority

When we sit down to meditate we should take the attitude that we are going to find out for ourselves just what is really going on with this consciousness thing, or this bit of sub-atomic physics, geology,

Meditation Advice

or whatever we happen to be contemplating. To sit down to meditate is to take a look for yourself at that from which all words and ideas flow. To sit down to meditate is to say your direct experience is fundamentally equivalent to any other human being that has sat down to do the same - not that your experiences will be the same, the mind seems infinite and non-local, but that your right to do so is equivalent. The authenticity you are able to bring to bear on the experience is no different in kind from that which any saint or prophet, thief or prostitute, highest sage or craziest denizen of bedlam, are also able to bring to bear. To sit is a radically democratic act.

What then of all those voices within us and within our society that insist we must give our inner allegiance to one of their belief systems?

It is one thing to learn from others, to be open to the hard earned wisdom of our foremothers and forefathers. It is another to believe someone other than yourself can provide any final truth for you or do the spirit work of your lifetime that only you can do. Your truth is hidden in a sacred place, it is inviolate. It lives between your innermost subjectivity and its relation to that which it perceives: the IS from which you came, from which you draw nourishment, and to which you will one day return. The gift of being alive is most intimately experienced as the depth of our mysterious subjectivity always and everywhere contemplating some objectivity, something rooted in earthly experience (however touched and distorted by imaginings it might be). We can freely bring our own faith to communal participation in the myths and rituals of our cultures. We need to celebrate and mark our dynamic dance on the edge of the unknown - but we cannot allow ourselves to lose ourselves in the crowd. By my lights that is what is meant by idolatry and is the mistake fundamentalism of any sort is guilty of.

Teachers are very different than preachers, though some preachers can be teachers if they know the score.

The powers that be serve their own purposes by confusing the distinction between authority and authoritarian. To respect authority and to be authoritarian are a pair of very different philosophical and psychological positions. By whose authority do those who torture and kill with impunity do so? By whose authority do those who resist those who torture and kill do so? Authoritarians insist that obedience is the highest of all virtues; the King, the General and the Pope will brook no disruptions in the ranks. Those who question authority wonder if what they are insisting upon is anything more than their own entitlement. Too often these leaders of mass movements and momentous decisions in the time of Homo Colossus, when pushed, act as though they ultimately own other people who must do as they say or be killed. Around such power structures other men gather to carry out the daily operations. The banality of evil Hannah Arendt found in Eichmann should ring ominously for us after the events of the totalitarian camps in Germany, Russia, and China. It is frightening that as a society we can become so callous to the things done in our name. Part of the haunting aspect of the banality of evil is how easily it seems a society becomes comfortable with the most outrageous things. Slavery and the Roman forum come to mind, filling the world with plastic garbage bags does too.

This authoritarianism has brought the earth to the brink of destruction. The oceans and forests are dying because so many small cuts are given her at the hands of those who were "just following orders."

Authority, on the other hand, derives from allowing authenticity to be the highest of all virtues, insisting that any human construct ultimately answers to a larger authority. We can say authoritarians locate the ultimate authority in human beings, though typically those human beings will claim divine sanction of one form or another. Those who respect authority but name authoritarianism a disease, locate relative authority in the earth and seek to work in harmony with the intelligence of nature's ways. The single most obvious aspect of the environment we find ourselves in is that it is ordered. Some people will study this order, learning as much as they can both from what was discovered about it before they came along, and seeking to add their own contribution to our understanding of

Meditation Advice

whatever their specialty happens to be. These people are an authority we rightly pay close attention to.

Sometimes these subject experts become convinced that they know what is ultimately best for you but in doing so they have crossed into the dictatorial territory of the authoritarian. 'It was for your own good' has joined the ranks of 'I was only following orders' in the informed moral conscience of the modern world. We have learned they are the preferred cover-ups for crimes of the worse kind.[21] We have learned, painfully, that such positions provide refuge for barbarisms. Missionaries with a sword are using god-words for reasons of political economy. No, Mr. Terrorist and Inquisitionist, you are not carrying out the work of "the Lord." What you do is sick and twisted. It does no one any good. It matters little if the dogma de jour is religious and the result is a crusade or if it is secular and the result is the gulags, they share a similar disease: someone else knows what is *ultimately* best for you.

At the end of George Orwell's 1984, when Winston is at the extremes of torture and counts four as five just before he dies, we are lead to understand that he has been made to love big brother. The point is that it was not enough to force the lips to say black is white, the authoritarian needs to change a person's very identity, *to own their subjectivity*.

This is the question the modern world, inheritor of so many horror-filled and blood drenched decades, cannot escape: how dangerous today are the true believers and how susceptible to demagogues do the people remain? Equally relevant, just how powerful are today's mind control techniques that cause these extreme personality changes and what, if anything, can individuals do about them?

Blessed are those who do not claim certainty, about that which no man can be certain. Blessed are those who do not doubt the real, that which no man can really doubt.

Imprints and the Child Within

The imprinting process is how the mind works. It creates loops, scripts, and sub-routines to aid its smooth operation. These imprints have neurological and biological underpinnings. Mediation is a reworking of circuits through the patient yet dependable process of strengthening those circuits we want strengthened, and starving those we do not think aid us in achieving the well lived life we are seeking. What fires often in the brain is strengthened and what does not will atrophy. This is neuroscience 101.

Many of these circuit imprints happened when we were young and our brains were still in their most active formation stage. Chance and circumstance delivered what now "feels" to us to be the most natural things in the world. Much of what we have learned serves us well but some things are problematic. We absorb the prejudices, for example, of those who raised us.

Folk wisdom has long recognized that childhood is "an impressionable age." Evidence from neuroscience shows how the developing brain adapts to the social and symbolic world it finds itself in, confirming what we have long suspected, namely, that there are imprints in the mind of an adult that were first laid down many decades previously. These imprints are found to be one of two kinds. The rule in the brain is: neurons that fire together, get wired together. We use this to learn things, knowing that the more we do something, the easier it will become to do. Language is picked up through repeated exposure and repeated practice. Anyone who listens to young children talking to themselves as they play together will hear the scripts strengthening these neural networks, repeating words again and again. Normal memory processing consists of these 'use it or lose it' kinds of cognitive processes.

The other family of imprints within the wetware of the brain consists of those singular experiences that were extremely intense for us personally. These traumatic events do not need repetition to form lasting imprints. They carve the plasticity of the developing brain in one shot, with one exposure. It is like burning computer code into a flash memory chip, the whole download happens all at once. The fact that this reorganization takes place immediately has consequences. Primarily, these networks are often left in isolation, without context. Normal memories are given repeated access which places them within the larger context of our lives. The traumatic event, in contrast, acts as though it is frozen in time. A war veteran suffering PTSD, for example, experiences their traumatic memories as if they were in some sense happening again, with their physiology showing responses to threat in the here and now. The other common symptoms include images that intrude themselves over everyday events, overlaying them with a sheen of horror. Doctors working with such people search for ways in which those memories can be defanged, so that they no longer re-traumatize their patients. Defanged they become re-contextualized as painful memories, making the world a bit sadder place perhaps, but no longer one haunted by shadows threatening to destroy the integrity of the self.

We can learn much from the suffering of others who may carry burdens much greater than our own. All people, not just those abused as children or returning from war, carry both kinds of these imprints. Our self-image is a reflection of the self we remember. The balance between healthy and toxic self-images these families of memories create differs greatly among people, and over a single person's lifetime. We are all, however, equally carriers of the burden of consciousness.

In bringing what our minds know to be true into our hearts, so that we feel the implications of our knowledge, we will encounter our own guardians on the thresholds. In Mindful Ecology it helps to remember why we are doing this. We are exploring the honest landscape of the human being, within and without, to honor the interdependent nature of the earth. We are doing this for a reason that is bigger than our individual lives. We will become more effective servants of the needs of the earth when we are able to penetrate the mystery of our own individual lives and no longer let the personas we have adopted rule us. To speak quite plainly: when we know the tragedy of our own story, it can reweave itself into the background of the larger tragedy unfolding all around us. Yes, we matter, what happened to us matters. Yes, you matter, what happened to you matters.

In the deepest reaches of meditation, compassion is extended to oneself and others. When extending compassion to oneself there will be times it touches on the child within to comfort and heal wounds from the past, to release the blockages of energy wrapped up in some of these imprints that are no longer serving the adult brain. When these encounters with one's own childhood take place they are accompanied by a variety of social and cultural image-sensations, including the religious where many of these conflicts originate. Follow the way that increases the amount of compassion with which you are viewing the child you were and are. Each of us, when we were young, exhibited the spontaneous, exuberant, innocent delight in being alive that is our birthright. It really is quite exciting to exist. It is the natural result of sensing the powerful biological flow of life within us and all around us. Even those with dark childhoods had some of these moments of harmony with that which is. Follow that. The same energy courses through your body today.

Here is a metaphor you might want to try on for size. The development of the mind of the human child was shaped over unimaginably vast spans of geological time. Pretend there are alien eyes looking out from the new born baby in your mind's eye; a being from deep time has arrived on earth; in the deep dark pools of her eyes you can still see the depths of black space, infinite in all directions. This is how you, too, started. You, too, have a unique tale to tell, irreplaceable in interdependence, important and necessary in the grand tale being told by your grandfather among the stars. Black-holes grace our pupils; we are the insides of the universe, its inside-outness; we are an infinite within to accompany the infinite without. This is what it means to be a sentient being. What else is there? What do you think you are?

Meditation Advice

7.
Mindful Ecology Contemplations

"We must uncenter our minds from ourselves;
We must unhumanize our views a little,
and become confident
As the rock and ocean that we were made from."
Carmel Point, Robinson Jeffers

Unhuman

This is from the poem in which Jeffers' idea of the unhuman, or inhumanism, was first introduced. The position is not misanthropic - but it is distinctly not human centered the way most of Western thought has uncritically and unconsciously assumed either. It is, as Jeffers said in the preface to The Double Axe, "the rejection of human solipsism and recognition of the trans-human magnificence." This is a very important point of view in Mindful Ecology.

The modern culture is a product of pervasive man-made environments. In addition to the physical infrastructure, which is fashioned to serve human needs alone, there is the mental landscape we also share and which is also no less oriented to exclusively human concerns. Here and there our built out environments touch on the lives of animals and plants but these too are put to work in zoos and ranches, farms and fisheries. There is nothing inherently wrong with our fascination with ourselves, after all, we are rather wondrously made but it can become as restrictive as a straitjacket if we fail to counterbalance its influence.

Contemplations

In the 1950s a California poet by the name of Robinson Jeffers developed a sensitivity to the all-pervasive nature of man-made artifacts and recommended a re-balancing by coming to appreciate what he termed the unhuman. He chose the somewhat awkward term primarily to defuse those critics which accused him of misanthropy. He was convinced mankind cannot find its meaning within itself alone, somewhat like Gödel's proof that no finite system can fully contain its own justification within itself. When we turn our wonder and attention to things not at all a part of the human realm, this does not mean we are anti-human. When our hearts beat to the rights and needs of the animal species we share this planet with, it does not mean we want to remove those rights and needs from humans but to extend what is obvious about our own subjectivity to the other sentient beings we find all around us. Animals are, in the final analysis, just like ourselves in so far as they are so obviously aware and moved by emotions.

The advances of affective neuroscience have put paid to the Cartesian notion that animals are different from us in kind. Mammalian brains include the same emotional neurological circuits as the human brain. By Occam's razor it stands to reason that we must accept that their experience of being alive feels to them not all that differently than it feels for us - with all that this implies for our relationships with the rest of the biosphere.

Some people object that while that may be true, the animals remain automatons because they do not possess the type of reflexive, self-aware consciousness that we experience. Logically this is hogwash. Philosophers of consciousness accept that there is no conclusive proof against solipsism; I might be the only self aware being in the universe and all other people could be fooling me. Reason, however, is a matter of what is most probable. My immediate experience is one of taking actions in pursuit of goals made urgent and desirable in my consciousness. Due to the actions of others exhibiting behavior patterns most succinctly explained by assuming they are also pursuing goals, it stands to reason their inner experiences must be not too dissimilar to mine. The fact that the DNA we share is almost identical further strengthens the logic of the assumption. Finally, in the case of human beings we are able to receive verbal reports from other people in which they describe what they are experiencing inside and find it correlates closely with our own experiences. All of these same bits of evidence apply equally to animals, particularly so to the mammalian species, except that last adaptation of word based languages. Animals do not speak with us about what they are experiencing inside. Not in words anyway. That animals communicate among themselves all the day long is obvious, so even that distinction is a bit shaky. These bits of evidence provide ways we moderns can begin to question our approach to the rest of the biosphere. We should remain aware though that the whole idea that other animal species need to be like us to have an intrinsic worth equal to that which we so freely assign ourselves, is exactly what an education in the unhuman is about overcoming.

We share many of the same biochemical cellular structures in our senses with numerous animals spanning evolutionary time. Eyes have been keeping an eye on the biosphere for literally time out of mind. We share many of the same neurological patterns and structures of numerous animals across evolutionary time as well, and these are the very structures involved in consciousness and memory. The aliens are among us, they are us, we are them, something like that. We need to shake ourselves up somehow to get a real sense for what is being pointed out here.

The difference between those who wake up each day in a dead, mechanical universe with an odd meaningless spark of awareness soon to be extinguished back into endless night and those who wake up each day in a living, organic universe of ceaseless fecundity thoroughly aware throughout seems to revolve around how one interprets consciousness. Lucky for us it is a subject we can investigate most directly. In fact, it is our personal destiny to do so. It is inescapable. It is also exactly what we are fundamentally studying in all our meditations.

All this mysterious conscious stuff, where did it arise from? What is the mother of mind? In the materialism of the West the answer is

Contemplations

assumed to be a wholly unconscious inanimate collection of matter. In the East however, as witnessed for example in the famous landscape art of China and Japan, there is recognition that even this inanimate environment is working with intelligence. The way of rivers and mountains, clouds and caves, trees and deep canyons, all under the rays of the sun and moon as the eons roll by, are shown weaving a tapestry of anything but random chaos. Here too, in the inanimate, there are endless forms most beautiful.

Much of what we have learned to regard as our own self is a social construct, but human society is not the be all and end all of our individual human experience. At birth and at our death, in orgasm and in pain, when the unexpected brings profound shock, in the most alien of our dreams and in countless other moments of consciousness other than the work-a-day version, we experience further dimensions of our being. Whatever else a human being might be, we are a member of the cosmic order and cannot fall out of its cradling of our every moment. It can be healing to remember that when the human order fails to maintain peace and sanity.

Intention, Our Light

It takes an enormous amount of courage to open ourselves to the emotional impact of our fears about the future. In a time like ours when corruption, lies and greed are in the driver's seats, thinking people carry rawness inside, a spot that is tender, painful. We who are ecologically critical of so many of our society's daily activities have had to stay quiet and get on with the necessary tasks at hand so many times we could not help but build walls of armor to protect ourselves. That is what happens when you take the abuse day after day while feeling powerless to alter course. Somehow we need to find a way to fight the numbness. Somehow we need to find the maturity that can thrive on the tension between the darkness we unwillingly participate in and the purity of our vision.

An epidemic of unhappiness seems to be spreading and not just because this was the week, as I write, that five of the world's major banks were declared criminal for manipulating foreign currencies and exchange rates. It seems to me that those I meet and talk with are running on empty. It is as if we are growing tired waiting for the next shoe to fall. Most people of my acquaintance are attuned to stories like the record breaking heat wave in India, the oil spill on the Santa Barbara coast and the goosing of our pretenses about regulating greenhouse gases by giving fracking endless capital and Shell a green light for deep water Arctic exploration, just to mention a few of the environmental stories of a typical week.

Our cultural stories have not left us well prepared for the most likely type of future bearing down on us and our children. In our stories, happiness accompanies material wealth with parties and good times being had by all. We lack stories about satisfying lives being found in challenging circumstances or stories that celebrate character for its own sake, even if it does not eventually lead to getting the girl, the house in the Hampton's and an eight figure bank account. Our stories are all crafted around the glow or glare of the spotlight; heroic deeds performed by larger than life gods and goddesses of the Silver Screen. No longer a slave to taste we bravely explore torture, blaspheme, and abuses of every kind within these same well-worn story tracks. Our stories are born from our sense of ourselves as a people. They dictate where meaning can be found, how relationships should go, what goals in life are worth pursuing and what each of us should expect from life in return.

Those are the expectations that are poisoning us, those that get deep inside and dictate to us what we should expect of life. For more and more people the expectations are not being met and the cognitive dissonance this is creating is coming to a boiling point. To escape being slave to your culturally created expectations – strengthen your intention. Ask not what your planet can do for you but what you can do for your planet.

Contemplations

The key to a mindful resistance to ecocide is to examine alternatives and ask what skills, attitudes and intelligence do we want to try and bring to the tasks of living well with our ecological knowledge and ethic? It is common enough that there is no viable alternative available today. That is when we need to respect the power of our intention. We may need to participate in a fossil fuel burning form of transportation to earn our daily bread but we do not have to approve of it. We can continue to foster in our hearts the desire to see a wiser society capable of meeting its transportation needs on a human scale. We may need to participate in the industrialized agribusiness to put that daily bread on the table but we do not have to approve of it. We can continue to foster the hope that sooner or later our societies will again live within their means and not depend on phantom acreage.

This may look to be a tiny thing in the face of the challenges we are confronting. Yet it alone might have the power to sustain the hard work of remaining open to our world and our times. It sustains the view that recognizes that in spite of all our ego-games and self-involvement there remains in us something that is pure, something that is clear. Our intention is beyond the limits of our cognitive mind since it includes our emotional makeup and our talents for navigating time. Our intention is not like a prayer or an aspiration, though it is often expressed in those ways. Our intention, if I was to put it into words, could be said to be the simple desire to see the end of unnecessary suffering for ourselves, our species and the whole of the living earth.

It seems such an outrageous dream, such an unrealistic hope. These objections miss the point. The path to the end of suffering is made up of steps that minimize suffering. Stands to reason, right? Those steps are surely within our reach. Not a day goes by where each and every one of us does not have at least a few opportunities to choose between lessening some form of suffering or not.

I prefaced this with a few words about our stories because it's easy to misunderstand talk about our pure intention. With our cultural stories for context, ending suffering is typically heard as 'nothing hurts' but that is not what is meant. We cannot remove the pain from life but we can remove non-essential suffering from that pain. Like many of the teachings related to the middle way what is being alluded to is subtle. The Stoics had some element of the right understanding when Seneca encouraged students with, "Most powerful is he who has himself in his own power." They recognized there is a value to staying true to one's ideals, that it delivers a happiness that is not dependent on the fickle winds of fate and fortune which we cannot control.

Those of us who dwell in the overdeveloped world are of necessity enmeshed in systems which harm the earth. Modern life entails participation in activities we don't approve of, activities actively damaging our planet. Often alternative means for procuring life's necessities are not available. Mass produced, mass marketed, mass consumed industrialized culture suffers from mono-vision. For all our talk of freedom and technological progress there is a surprising dearth of real choices for how we work, move, eat and educate ourselves. Monopolies abound, dissensus not so much. An honest appraisal of our situation recognizes that there is much we as individuals do not control.

Yet, it is equally true that there is very little we as individuals do not influence in any fashion. Here the cracks begin, the cracks where the light gets in. Here is why holding one's intention clearly is so important. Knowing what you stand for both steadies us for the hard work of remaining open and readies us for taking advantage of any opportunities that present themselves to participate in more life affirming alternatives.

Ultimately the core industrial processes as we know them will prove to be a short chapter in our species history. Consuming and wasting as many non-renewable resources as fast as possible to maximize profit and growth is simply not a sustainable value system for organizing cultures. The temporary energy bonanza now coming to an end enabled it and we were quick to add the delusions about our special place in the sun it required. Today as the age

Contemplations

of consequences is just getting started, the search for alternative values and stories by which to organize and understand our social lives is apparent everywhere. We are losing our reference points and along with them the legitimacy of our former institutions. It can be very unsettling to live through the twilight of idols.

Get to know your pure intention. It is not the weakling modern ad-copy makes it out to be. It does not guarantee 15 minutes of fame, nor riches, nor even popularity. It will, however, provide a steady light – just that which is most valuable in a time of darkness.

Molecular World of Deep Time

The contemplations of Mindful Ecology start and end from a particular point of view or context. That context is larger than that which we have been socialized to or educated in because it includes the reaches of deep time as revealed by biological evolution and the reaches of deep space as revealed by the evolution of the elements in the cosmos starry nurseries. The difficulty in communicating this view correctly is that it seems to point to states of mind and intuitions that are way out there, as the expression has it. In fact, the way out - is down. Everything about these contemplations is involved in a re-dedication of ourselves to earth and her needs. We greet these needs first in ourselves as we attend to the cry inside over the ignorant destruction of our precious home. We continue to attend to these needs as we listen for the song of meaning in the soft breeze of Eden's eternal morning, and in the din of the hurricanes coming to slap us awake.

This is one way of thinking about things some might find useful in learning to "think like a mountain," as Aldo Leopold taught us to do.[22] It contrasts human "evil" with natural "evil" and concludes we suffer from exaggerated fears about the reality of our existence. Learning to rest in that reality allows us to live each day well. There are a number of alternative means to guide the mind into the same view. Each of us finds their own most effective ways for ourselves eventually and even those, in my experience, do not remain unchanging. The point is not how we get there but where it is we arrive. We are learning how to get in touch with the identity we carry on the elemental planes, where the molecular dance of DNA's immortality enlivens our own hearts and minds. We are learning to start and end our every contemplation in this 'peace which surpasses all understanding.' The molecular world features subjectivity, the two are inseparable in our actual experience. This is what we embrace as our birthright. In doing so we overcome the fear that cuts us off from a full-hearted acceptance of our life as a man or woman of earth.

We are looking to turn our attention to the many gifts our nature provides us. The populations of bacteria in our guts that allow us to absorb nutrients from the food we eat, for example, might seem a humble addition to one's body but seen rightly play a much more fundamental role in sustaining your life than hands or feet or sexual organs. It matters not whether we imagine the source of these gifts, this pervasive intelligence we find throughout the universe at any scale on which we care to look, as personal or impersonal. The facts indicate that it "cares" to sustain that which finds its most fundamental expression and experience in personal subjectivity. The living universe, after a fashion, looks after us. That this is not at all a way to sneak in the theological god of childhood faith through the ecological back door will, hopefully, become quite clear as we proceed. There are no rhetorical flourishes here to deny the problem of evil.

People of good will look at the horrors of recent history and ask themselves what can one person possibly do to minimize the chances that such evil is again allowed to be unleashed. The lessons of Dachau, Dresden and Nagasaki weigh heavily on every person who learns about them, particularly our children. It is easy for us to forget how devastating it was when we were first exposed to the facts about the modern world's recent history. Though one part of our mind might just file these things away as just so many other facts, there is no escaping the emotional impact these things also have. Acts of such indiscriminate murder, made giant through the power of industry, call into question the very existence of nobility and dignity in man. We fear that these are the things that reveal the real truth about ourselves. The emotional impact questions the very things that bring meaning to human lives; concern and love for particular individuals. It is frightening that human beings can be driven to such acts of extreme depravity. They hurt us not just in our flesh but in our soul.

These types of great evil always entail the sacrifice of blood but it hungers for something more; the cruelty that drives the men controlling and ordering such murderous plots wants to crush the last bit of dignity it can find in human subjectivity. Put plainly: there is an evil hunger to degrade that which is seen as innocent or weak. The degradation of women and children through violence and abuse is the long un-discussed underbelly of our societies, though the pervasiveness of gonzo pornography is now rubbing our faces in it. Haunted by a rape culture just under the surface, all of us must confront the outcome of this me-first selfishness that forms the core ethic of our economics.

The sadist needs to deny the reality of the subjectivity of those around him. We fear the universe is an uncaring, blind, machine designed to inflict pain. Our science seemed to point to life as little more than a chance mistake, a happy accident but nothing that really means anything to the universe in the long run. Though it might seem the myriad animals all around us are as sentient as we ourselves are, Descartes taught us to see that this was just an illusion. Animals became, for modern man, little more than machines with eyes. We have done with them whatever we wish, accordingly. As the molecular insights of biochemistry accumulated it became more and more difficult to draw a line between mankind and these animal machines. It was a slippery slope to start on, denying the subjectivity that was so evident all around us throughout the animal kingdom. Our own subjectivity was targeted not long after. The vicious beast we fear we are, our semi-conscious story went, is not our fault for we are nothing more than robots, automatons in the hands of molecular determinism. The sadist acts out this logic, traumatizing other living beings in a defiant refusal to recognize their legitimate claim to exist. The sadist has taken it upon themselves to dispense with the worth of another's identity, their target's essential existence.

It is this point of view that poisons. The dead universe imagined by early Newtonian physics has no relation to the universe we actually encounter. It was a simplification, a model, to give our minds purchase on the complexity of the molecular universe. The unity of celestial and terrestrial mechanics in gravitation's terms of mass and momentum blazed the trail into the mathematically modeled scientific knowledge that has become so dense and abundant today. So far so good. We looked to nature in all its organic complexity and found we could tease out singular functions and model them in our machines. Though this reductionism captures the essential relationships involved it does so in a very fragile medium. Unlike the organic inspiration for these inventions of ours, our machines do not reproduce themselves and are singularly incapable of adapting to changing circumstances. The success of reductionism caused us to get our bearings turned around all topsy-turvy. We began to consider organic living things as consisting of no more than levers, valves and winds instead of recognizing these mechanical features played no more than supporting roles in the overall work of living; namely, the self-reproduction by which life continues, and the consistent adaptation to the changing environment homeostasis requires by which a life continues.

Granting subjectivity as a primary characteristic of the universe exposes the immense error the sadist has made. Already each individual knows that their own awareness, their own immediate experience of sentience, is undeniably real. To justify hurting others and taking from them their very right to exist the sadist has had to deny that this self-evident awareness also exists in anyone else, or if it does that it has an equal right to do so. If the universe is not a monstrous machine designed to inflict pain, if it is more like a living

Contemplations

mother and father to all that exists, then the monstrous machines some men have designed to inflict pain stand alone as examples of existence built on delusion and confusion. Ethnologists assure us the only cruelty we find in the animal kingdom is that which is involved in eating and other direct survival needs. The added cruelty human sadism is uniquely able to bring into existence cannot be blamed on our evolutionary roots. Though it has benefited the military industrial complex to propagate this view that humankind is helplessly and fundamentally flawed and destined to use their wares, it does not make it so. That human beings must remain enslaved to sadism and sadists is propaganda from the largest part of humankind's giant industrialization, the violent heart of Homo Colossus.

The point of contemplation in Mindful Ecology is to strengthen us to be present with the reality of our world today. This necessarily entails making a place in our minds for the sadism such as we saw displayed in Dachau, Dresden and Nagasaki. That is a place occupied by humankind. The rest of creation lacks that demonic depravity.

Out in the real world, beyond whatever man-made constructs might be in our internal and external environments, we have a chance to encounter the original purity which first inspired the many creation stories to shout the ultimate affirmation of existence, each in their own way: "It is Good!" The proper ecological perspective would take what we have learned about the rest of nature seriously. In nature there is eating and there is the compost pile. Like the Christian church fathers used to say about story of crucifixion and resurrection that the Christ rescued death from the grip of hell and the devil. The ecological view also removes the natural process of living and dying from the grotesque forms invented by sick and twisted human beings. Accepting the reality of reality entails accepting that life entails eating life. In all its countless forms this is just 'regular old death.' In the natural process all living things give back when it is their turn to become food for others. That process of becoming food for others is how life both makes room for its new generations and how those new generations are ultimately fed. Massive cycles of molecules are sewn into such exquisite patterns for all this to happen. The correct point of view recognizes that everything that is most precious to oneself, as this mystery of awareness has unfolded uniquely in your own subjectivity, was given you by that which created the totality of these patterns.

We need to learn to see more clearly the truth that we have our roots in the mountains and streams, clouds and oceans, not only in the precious, yet ultimately tiny world of human affairs. Subjectivity teaches us to value the few precious individuals whom we have come to know and love above the comings and goings of the masses. This teaches each of us that the most important parts of the human experience have never been captured in our history books. The real action will not take place tomorrow, and it did not take place in the past. The real action is always right now where we encounter the mystery of it all. Here is where the real unfolding of a life takes place, here where the whole organism encounters the whole environment. In this gestalt there is no separation of conscious feeling from physiological emotion or either from intellectual thought. Humans are big, too big to be wholly contained in the cities of civilization driven by armies of bureaucrats. We are more at home under the sky than under a roof. This is something too many modern people have forgotten. We need the sky. We need the sky for we are large.

That which calls us forth, sustains us, and dissolves us does not have the same nightmare level of cruelty it is possible to experience at the hands of man. We fear the universe is as cruel as sadistic human beings. I submit that there is no evidence for such a conclusion. This is our imaginations running wild in the worst of anthropomorphisms. Death is not the ultimate sadist.

When the artificial fear of traumatically unjust suffering induced by human perversions is removed from the breast, we find we are more at home in the world just as we find it. It is not our death we need to fear; it is not our desires, nor our needs for love and attention; it is not even the painful tortures twisted minds are driven to inflict on others. We do not even need to fear the end result of all our crazy disrespect for the living planet from which we draw our every sustaining breath. When the mind gets this ecological view, there is simply nothing to fear with the kind of fear that we had been taught was appropriate for our human circumstances. It is not.

We learn to sit like a mountain.

Life is a Door

All of space surrounds us at every moment. Like a fish that does not recognize the water it swims in, so we do not remain conscious of that which ceaselessly embraces us. This awareness of emptiness is but one of a myriad levers within the landscape of the mind that is capable of reawakening pure awareness within us. These levers are able to open a door of perception through which we can walk to find ourselves in a world made a-new, the sacred world.

Earth is not a dead thing. It is not blind, dumb, and deaf, nor is it without feeling and awareness. A culture dead set on using and abusing the earth had to tell itself these things were true but it did not make them so. Just as the Native American elders warned us, we seem to be blind to the life of the land and deaf to its song. How else could we do the things we do?

This is not a moral issue, or at least, not only a moral issue. It is a matter of perception, a matter of how we see ourselves in the big scheme of things and how we see our home. There are medicines to heal what ails us. We are becoming a terminal case. Quick, open the heart and listen to what the land wants.

This jewel, spinning silently in space, radiates with the luminosity of its biosphere. Here is a nexus on which the convolution of complexity reaches to the evolution of the brain. The awareness embodied radiates out to the farthest reaches of infinite space. Is it not true that in some puzzling way all the stars are on the inside? What else does it mean to perceive what we perceive?

Every day when you wake up the earth reaches out to embrace you, to cradle your awareness and provide the ground on which to carry out the acts of will for which you were born. Everywhere this union of your body-mind with the planet is an embrace filled with life. Everywhere you look you find flowing exchanges of energy and information that go back endlessly through beginningless time.

We can define this part by using this other part and so knowledge and understanding can extend its reach throughout the whole of a transparent universe. What we cannot do is step out, or fall out, of this embrace. There is no there there in those imagined heavens far beyond the earth life we know. There is the content of the moment. It is real but it is fleeting. Quick, open the heart and listen to what the land wants. Soon your grandchildren will be walking where you are walking right now. What will your blood be teaching them? What will your bones be making available for them?

We are prone to distrust the voices from the future we hear in our dreams. Perhaps then we can break through the cocoon of callousness by speaking with our dead. So many lives have come before us. Who crafted your bones and shaped your eyes? Whose passions stir your blood? Those interest of yours and driving obsessions, just how sure are you that they are yours alone and not the continuation of a playing out of desire pure enough to reach through centuries with ease? We celebrate tragic love in Romeo and Juliet but what of the lovers at play? Isn't the play what the whole human race is an elaboration of? Is not the human embrace, both when all the lights are on and when it is the warming ember comforting us in the darkness, perfected in bliss? This is who we are. In our hearts the youthful teenagers in first sexual union, exploring all the world through the doorway of skin, ceaselessly celebrate the unborn of our future, with a laugh and a tear, with the orgasmic sigh and the sweet little death.

Our husks have become too dense, too rude and course. The inability of our culture to integrate compassion throughout our practical affairs must be overcome. We must learn not to fear our emotions. It is scary to open up to the sacred world, to allow a full

Contemplations

heartfelt recognition of just how much you love life and the lives of the people you have known. As fleeting as a bubble, not more than a lightning flash... how can we feel so deeply? The luminosity of those in the embrace threatens to rage out of control if its basic innocence is lost. Then we enter the dark heart where the dark arts of forced sex and the prostitution of 'souls' take place.

We are finding our way, a bit further with every generation, to an understanding of the real strengths and weaknesses of our place in the bigger scheme of things. There is nothing like excess to show you where the boundaries are. Today ecosystems all over the earth are revealing exactly where we trespass those boundaries. The recognition of our ignorance rebounds as wisdom; it was an inability to see the dignity of all that lives that poisoned our culture. The spear we sent into the animal world - the factory slaughterhouses and the sixth extinction - is coming back at us; we see it everywhere; now the dignity of the human being is under attack.

Incapable of celebrating our shamanistic roots, we cannot even begin to talk about the states of mind that fill our lives with a type of cosmic glory. What we have to offer our children comes from a poverty mind, one filled with fear. Have you not heard? The gods, whatever they might be, depend in some way on the mind of man. We, by our being, allow them to have their being. Ring the bell, wake the children, the days of celebration are here. Why? Because you have, right now, a precious human life. It will not always be so on the planet. What will you do with today? Will you spend it working for meaningless ends?

Quick, open the heart and listen to what the land wants. Who cares what other people think, if what they think is killing everything?

Sometimes only art is capable of saying what needs to be said, be it a painting by a master such as *Pregnancy* by Alex Grey or stumbling poetic prose by yours truly:

Those ancient ageless young ones laughing in the golden sun and whispering the sweets nothings of eternity into one another's ears under the moonlight, your mother and father from the stars long ago enthroned on earth in an age lost to deep time - they are as close as your heart. You are their beloved child, the hope which was longed for in the most intimate dreams of countless millions. They call you by name and offer you a task in time, if you will. Every day in every way remember that on some other level this is what is really real about emergence in evolutionary spacetime; this deathlessness and love. The Universe spun out from ceaselessly shuffling the periodic table was not enough to contain the primeval shout of creation's bursting will to exist. It required the green fire of life itself, aware and vulnerable; it required individuals because it required the capacity to love. The contained celestial fire we see in one another's eyes is the proof that it was here, right here on earth, that the skill managing the elements throughout countless galaxies has perfected the expression of graceful care.

Space Age - Earth Age

Hubris drove the protagonists of the ancient Greek tragedies to trespass into realms reserved for the gods and those so cursed would soon pay the price. Has Homo Colossus trespassed into sacred territory?

When humankind split the atom we unbound that which god had bound, as it were. In this rupture of the weak and strong forces, the hand of man brought a rupture to the very glue which holds together existence itself. These violent explosive forces, capable of tearing at the very fabric of our universe, existed nowhere among the earth's many processes until we brought them forth through the use of our technology. Perhaps, if nuclear war descends on us or our descendants, we would only consider our voyage into the realm of the sub-atomic a trespass into the realm of a very angry god indeed. It certainly has not delivered energy "too cheap to meter" as its early supporters once promised.

Contemplations

When humankind pierced the thin blue line of the biosphere with its rocket to the moon, perhaps, once again we trespassed. We once thought our technology would lift our species out to the stars, though modern knowledge forces us to admit only much more modest achievements will ever be possible. Like an unruly adolescent stepping out for the very first time, we shot our tin can into the sky and threatened never to look back. We found a dead moon, dusty with the un-breathing of cold space. Giant industrialization had created the final prosthetic, one capable of overcoming the resistance of the planetary gravity well. Homo Colossus threw Homo Sapiens out into the starry expanse, displacing us. Be careful what you wish for.

Being mindful of ecology, trying to take in the long view, it could very well be that our outrageous dedication to science and technology that allowed these trespasses just might be our saving grace. Both of these trespasses may have been necessary. Like cybernetic steering they are the warp and woof that might keep us on track somewhere in the middle way between the extremes they represent. There is no success like excess. In our steering into the future our excesses were both brought about through the one time possibilities oil-fueled Homo Colossus provided. The largest machine ever built, CERN, pierced the realm of the small far enough to find the Higgs boson. The largest corporate structure ever built, the modern military industrial complex, was needed to put a man on the moon. Too tiny, too big, just right. The tale of industrialized civilization's giantism is one of trespassing limits in a Goldylocks universe.

One can hope it taught us our place. If this turns out to be a lasting result of our excesses - once burnt twice shy - then I, for one, am almost willing to believe our species' one time shot at oil-fueled giantism was well spent. The ingenuity and cleverness brought to our Faustian deal with the devil to become masters of our world paid off, in a typically devilishly cursed way. It just might turn out that all along we were governed by a wisdom that could not be thwarted despite our petty greed and ignorance. That will prove to have been the case if we can distill the lessons to be learned from the mistakes we have made. Then we can reasonably hope for a world in which our children's children live as true masters of their technology, because they have mastered themselves.

Human beings were taught by Homo Colossus to expect the future of their species to be one that unfolds out among the stars. High technology was going to shrink the distance between star systems just as wireless telegraph shrunk the distance between continents. We were entering the space age.

This was delusional thinking on our part. Homo Colossus is unsustainable. Without the massive energy and material inputs it requires it will die. This was the betrayal of technology in the small; it proved unable to provide a limitless source of clean energy to substitute for oil. This was the betrayal of technology in the large; there is no space program without Homo Colossus, no outer frontiers beckon with a promise of a new and better world.

What this means is nothing less than that we get to re-think the whole point of what we have learned and what we have done and, importantly, what we will choose to do tomorrow. The human footprint on the moon made it real for us that a planetoid object could be as dead as dust. We could not help but wonder about the footprints we were leaving on the earth. In our own ecological fears that image found a deep resonance, almost like a catalyst to drive home the point that we really could blow it down here. The Hubble Telescope's countless galaxies (galaxies!) has, arguably, reawakened our wonder but it has certainly forced us to grapple with just how far the distances between the stars really are and how vast that which dwells above our heads actually is in all that it encompasses. These are iconic, mythic, archetypal images of lessons not soon to be forgotten. They may well continue to play iconic roles for our ancestors long after Homo Colossus has died. They are not, however, the images I believe will prove to have been the most critical. Like a holy angel watching over us in our trespasses, it is the image of the earth as seen from space that holds our truth.

It is one. It is rare. It is threatened.

Contemplations

We did not enter the space age, we entered the earth age.

And coming home,
after journeys long and lost among satanic mills,
they saw the organic earth as if for the first time;
precious,
ringed in white and blue amidst the infinite black;
majestic.

We are learning to see her as she really is.
Sacred.

8.
Perspective and Peace of Mind

Think like a Mountain

Hope.

That is what is currently in short supply. This is an error born from looking to our own devices to solve issues of suffering that are far beyond our powers to deal with. "Give me the wisdom to accept the things I cannot change, the courage to change the things I can, and the wisdom to know the difference," Reinhold Niebuhr's Serenity Prayer from the 1930s teaches. We are embodied beings only able to affect those things we can grasp with our hands (however enhanced our prosthetics) or touch with the expressions of our arts. There is a rich vein of happiness to be found in the simple things of our earth, for this is where the non-dramatic yet vital processes are serenely going about their business, working to support life and assuring that the rain will continue to fall on the just and the unjust alike.

This is a precious garden planet on which we are lucky to be alive. The great events of the day that seem so important and pressing, they can be held more gently. It is not an avoidance of responsibility in blissful hours of meditation that this work is all about, not at all. It is about finding the inner strength to wield your will joyfully. Do what you can and you will find in fulfilling your deepest values that somewhere along the way you learned to be happy. The trap is to think the little, simple things that you can actually do are meaningless. Some part of us revolts against the limitations of our creaturehood and desires to be a god capable of ordering all things right. If that is your approach, that you are a failure every time you wake up and the world is not lots better yet, no wonder you are depressed.

Peace of Mind

You are setting yourself up for failure. This is the confusion about what we can and cannot change that the serenity prayer is asking a higher power to heal us from.

The pain of earth's groans right now is frighteningly real for us who are sensitive to it. We who are being called in this way need to be particularly wary of allowing exaggerations of the poisons and trigger points to unsettle our proper reasoning. The DNA is dancing us through deep time, and we can be quite comfortable trusting its inner wisdom. We need to learn to *feel* how amazing its accomplishments truly are.

All that has ever really mattered are the little, simple things. These are what the fossil records in the mountains teach us. This is what our meditative contemplations on our own bones, our longest lasting material expression on this earth, show so clearly; for what are these fossilizations but the record of little, simple things? It is what the human hand can hold that holds the diamond of infinite worth. 'Put your hand to the plow and do not look back' the Bible counsels. I've seen it work for others and am seeing it work in my own life, how it carries me more lightly when I focus on the small "k" kingdom I am able to build than when I focus on the big "K" Kingdom of Earth I wish I could aid.

Let that which lifted the mountain ranges be that which lifts mountain ranges, and let that which runs and plays on the mountain for a season or two, be that which runs and plays. Human beings might well be, at least potentially, the most noble creature in all the universe. Who has plumbed the full depths of love and virtue? We children of the stars are powerful when we play. Never lose hope; never believe that the power of the child at play cannot overcome war in the hearts of all; never believe that the value of kindness is a fluke in this universe – and rest in that; be quick to say thank you and you will find, eventually, the courage to fully say yes to your destiny. There is a reason earth is calling our hearts, each and everyone of us who shed tears for her. This I truly believe.

Notes

1. Hans Rosling asserts that most people hold to an overly pessimistic view of the world. He believes it is due, in part, to our lacking an information rich appreciation of just how successful our scientific, engineering, and social achievements are. In *Factfullness: Ten Reasons We're Wrong About the World - and Why Things Are Better Than You Think* he provides a useful correction for those who might despair that there is nothing but bad news concerning modern global affairs.

2. Buddhist Economics was said to have first been practiced in the reign of the Indian Buddhist Emperor Ashoka (c. 268 to 232 BCE). The term was popularized in the West when an essay of that name was included in E.F.Schumaker's influential 1973 *Small is Beautiful: A Study of Economics as if People Mattered*. Buddhist Economics studies what benefits and harms come from various economic activities in order to identify which lifestyles are conductive to self-actualization and environmental harmony. Today there are many hopeful developments along these lines including Bhutan's use of an index they developed for tracking their Gross National Happiness (GNH) in place of the common Gross Domestic Product (GDP). An exemplary scientific journal very relevant for the ecologically concerned and aligned with the spirit of Buddhist Economics is published online by Springer, *BioPhysical Economics and Resource Quality*, https://link.springer.com/journal/41247

3. *The Secret of the Golden Flower, A Chinese Book of Life*, pg 128. Dr. Jung further elaborates, "It is my firm intention to bring into the daylight of psychological understandings, things which have a metaphysical sound, and to do my best to prevent the public from believing in obscure power words." I think this should inspire our work in Mindful Ecology as well.

4. This is from a sacred song sung by H.H. The Karmapa entitled *An Aspiration for the World*. It is available on the Mindful Ecology website.

5. The thermodynamic qualities of biological science are elaborated in quantum physicist Erwin Schrodinger's lectures published in 1944 as *What is Life? The Physical Aspect of the Living Cell*. See also Eugene Odum's use of thermodynamic analysis found throughout his critically acclaimed and highly recommended textbook treatment of ecological principals *Fundamentals of Ecology* (1971 Third Edition my preference).

6. Catton, *Overshoot*, pg. 24

7. *Proverbs* 29.18

8. *Five Years*, written and recorded by David Bowie, released 1972, RCA Records

9. Beat Godfather Meets Glitter Mainman: William Burroughs Interviews David Bowie, *Rolling Stone Magazine*, Feb 1974. Accessed 5/2018, https://www.rollingstone.com/music/news/beat-godfather-meets-glitter-mainman-19740228

10. *Where Are We Now?*, written and recorded by David Bowie, released by 2013, ISO Columbia

11. *What a Wonderful World*, written by Bob Thiele and George David Weiss, recorded by Louis Armstrong 1967, ABC Records. You Tube has the version with a spoken introduction that is very relevant for those pained by ecological facts.

12. *Overshoot* pg. 219. A darkly prophetic page if there ever was one.

13. Population controls 'will not solve the issue,' *BBC News*. Accessed May 2018, http://www.bbc.com/news/science-environment-29788754

14. Turner, Graham. (2008). A comparison of The Limits to Growth with 30 years of reality. *Global Environmental Change*, 18, 397-411. 10.1016/j.gloenvcha.2008.05.001.

15. Smithsonian Website, accessed May 2018. https://www.smithsonianmag.com/science-nature/looking-back-on-the-limits-of-growth-125269840/

16. See Oreskes and Conway, 2010, *Merchants of Doubt: How a Handful of Scientists Obscured the Truth on Issues from Tobacco Smoke to Global Warming*

17. "The menu is not the meal," insisted Alfred Korbensky in the 1930s. His *Science and Sanity, An Introduction to Non-Aristolian Systems and General Semantics* remains a fascinating and timely attempt to utilize our scientific knowledge, and the scientific approach, to better our daily lives in a very practical way by watching how we use language. Mindful Ecology follows in that spirit. Most of our social failure to deal effectively with the accelerating ecological crisis is due to politicians and thought leaders reacting to the words used instead of the reality the words are pointing to. This makes societies feel powerful, as though they were in control where they are not. It is a seductive hubris to think that reality is as easy to manipulate as the words we are using to describe it.

18. Goldhammer, *Under the Influence*, pg. 198, italics in original.

19. Lifton, *Thought Reform*, pg. 420.

20. The term stinking thinking is used in cognitive therapies. These identify self-defeating errors in how we think about ourselves and our world and challenge them on the spot. People concerned about the collapse of societies certainly need to be on the lookout for exaggerations, black and white thinking, and other cognitive errors the human mind is prone to. Dr. Albert Ellis and Dr. Aaron Beck have written extensively in this field and are recommended.

21. In *For Your Own Good: Hidden Cruelty in Child-Rearing and the Roots of Violence* Alice Miller lays out ramifications of child abuse that are difficult to ponder but important to understand in our time of rising school shootings and questionable populist leaders.

22. In *Sketches Here and There* Aldo Leopold introduced the idea of appreciating the inter-connectedness of ecological systems on their own terms in a very personally moving way. In an essay entitled 'thinking like a mountain' he described learning something new the day he watched the green fire of life die in the eyes of a wolf he had shot. It became an iconic image of a human being learning a new respect for the sacred depths of earth's mysterious molecular ways.

Bibliography

Bateson, Gregory, and Alfonso Montuori. *Mind and Nature: A Necessary Unity*. Cresskill, N.J: Hampton Press, 2002.

Carroll, Sean B., Jennifer K. Grenier, and Scott D. Weatherbee. *From DNA to Diversity: Molecular Genetics and the Evolution of Animal Design*. 1 edition. Malden, Mass: Wiley-Blackwell, 2000.

Catton, William R. *Overshoot: The Ecological Basis of Revolutionary Change*. Reprint edition. Urbana; Chicago: University of Illinois Press, 1982.

Coffin, Arthur B. *Robinson Jeffers; Poet of Inhumanism*. 1st edition. Madison: University of Wisconsin Press, 1971.

Dawkins, Richard. *The Extended Phenotype: The Long Reach of the Gene*. Revised edition. Oxford ; New York: Oxford University Press, 1999.

Edwards, Paul N. *A Vast Machine: Computer Models, Climate Data, and the Politics of Global Warming*. First Edition. Cambridge, Mass: The MIT Press, 2010.

Goldhammer, John. *Under the Influence: The Destructive Effects of Group Dynamics*. Amherst, N.Y: Prometheus Books, 1996.

Greer, John Michael. *The Long Descent: A User's Guide to the End of the Industrial Age*. 1st Edition edition. Gabriola Island: New Society Publishers, 2008.

Grey, Alex. *Sacred Mirrors: The Visionary Art of Alex Grey*. Inner Traditions International, 1990.

Hall, Charles A. S., and Kent A. Klitgaard. *Energy and the Wealth of Nations: Understanding the Biophysical Economy*. 2012 edition. New York, NY: Springer, 2011.

Heinberg, Richard. *The Party's Over: Oil, War and the Fate of Industrial Societies*. 2nd edition. Gabriola Island, BC: New Society Publishers, 2005.

Mindful Ecology

J.A. Bayona. *A Monster Calls* (DVD). Universal Pictures Home Entertainment, 2017.

Jeffers, Robinson, and Tim Hunt. *The Selected Poetry of Robinson Jeffers*. Stanford, Calif: Stanford University Press, 2001.

Jensen, Derrick. *A Language Older than Words*. White River Junction, Vt: Chelsea Green Pub. Co, 2004.

———. *The Culture of Make Believe*. Second Printing edition. White River Junction, Vt.: Chelsea Green Publishing, 2004.

Jones, Alexander (Editor). *The Jerusalem Bible, Reader's Edition*. Doubleday & Company, Inc., 1968.

Jung, Carl Gustav. *Modern Man in Search of a Soul*. Harcourt, Brace & World, Inc., 1965.

Korzybski, Alfred. *Science and Sanity: An Introduction to Non-Aristotelian Systems and General Semantics*. 5. ed., 3. print. Brooklyn, N.Y: Inst. of General Semantics, 2005.

Kunstler, James Howard. *The Long Emergency: Surviving the End of Oil, Climate Change, and Other Converging Catastrophes of the Twenty-First Century*. 1st Edition edition. New York, NY: Grove Press, 2006.

Leopold, Aldo. *A Sand County Almanac and Sketches Here and There*. New York usw: OUP, 1972.

Lifton, Robert Jay. *Thought Reform and the Psychology of Totalism: A Study Of "brainwashing" in China*. Martino Publishing, 2014.

Lü, Dongbin, Richard Wilhelm, and C. G Jung. *The Secret of the Golden Flower: A Chinese Book of Life*. London: Routledge & K. Paul, 1962.

Meadows, Donella H., Jorgen Randers, and Dennis L. Meadows. *Limits to Growth: The 30-Year Update*. 3 edition. White River Junction, Vt: Chelsea Green Publishing, 2004.

Miller, Alice. *For Your Own Good: Hidden Cruelty in Child-Rearing and the Roots of Violence*. Translated by Hildegarde Hannum and

Hunter Hannum. 3rd edition. New York: Farrar, Straus and Giroux, 1990.

Mumford, Lewis. *The Myth of the Machine*. San Diego: Harcourt Brace Jovanovich, 1970.

———. *The Myth of the Machine. 2: The Pentagon of Power*. New York: Harcourt Brace Jovanovich, 1970.

Naess, Arne. *The Ecology of Wisdom: Writings by Arne Naess*. Edited by Alan Drengson and Bill Devall. Berkeley, CA: Counterpoint, 2008.

Odum, Eugene, and Gary W. Barrett. *Fundamentals of Ecology*. 5 edition. Belmont, CA: Cengage Learning, 2004.

Oreskes, Naomi, and Erik M. Conway. *Merchants of Doubt: How a Handful of Scientists Obscured the Truth on Issues from Tobacco Smoke to Global Warming*. 1st U.S. ed. New York: Bloomsbury Press, 2010.

Orwell, George. *1984*. New York, NY: Signet Classic, 2002.

Richard Curtis. *Love Actually* (DVD). Universal Pictures Home Entertainment, 2009.

Rosling, Hans, Ola Rosling, and Anna Rosling Rönnlund. *Factfulness: Ten Reasons We're Wrong about the World--and Why Things Are Better than You Think*. First edition. New York: Flatiron Books, 2018.

Schrödinger, Erwin. *What Is Life? The Physical Aspect of the Living Cell*. Reprint. Cambridge: Cambridge Univ. Press, 1983.

Schumacher, E. F., and Peter N. Gillingham. *Good Work*. New York: Harpercollins, 1980.

Schumacher, Ernst F. *Small Is Beautiful: Economics as If People Mattered*. New York u.a: Harper [and] Row, 1975.

Sussex, University Of. *Models of Doom: A Critique of the Limits to Growth*. Edited by H. S. D. Cole and K. L. R. Pavitt. First American Edition edition. New York: Universe Pub, 1973.

Tainter, Joseph A. *The Collapse of Complex Societies*. 23. print. New Studies in Archaeology. Cambridge: Cambridge Univ. Press, 2011.

Timothy Scott Bennett. *What a Way to Go: Life at the End of Empire* (DVD). VisionQuest Pictures, 2007.

CPSIA information can be obtained
at www.ICGtesting.com
Printed in the USA
FFHW010953020219
50343185-55420FF